HELM&
HORIZON

Daily Leadership Principles for the Motivated Sailor

STEVEN-PAUL LAPID, USN (RET.)

SHINE
PRESS

Jodi K Costa, LLC
Tampa, Florida

SHINE PRESS

Copyright © 2025 STEVEN-PAUL LAPID
HELM & HORIZON
All Rights Reserved. Printed in the U.S.A.

Publisher: Shine Press
4522 W. Village Dr. #1294
Tampa, Florida 34624
Shine-Press.com | Jodi@Shine-Press.com
Shine Press is an imprint of Jodi K Costa, LLC.

For speaking engagements, event invitations, bulk orders, and other author requests, please contact the publisher: Jodi@Shine-Press.com

Line Editing by: Elliott San, Good Marrow Inc.
Book and Cover Design and Formatting: Shine Press

Paperback ISBN: 979-8-9993665-8-0
Ebook is also available.

FIRST EDITION

To Michelle:
You are my compass and steady harbor.
Thank you for your patience, your strength,
and your unwavering belief in this mission.

To Maliyah, Maya, Michael-James, and Maddox:
You are the future generation of world-changing leaders.
Wherever your rudder takes you,
may you lead with courage, conviction, and heart.

CONTENTS

FOREWORD

Leadership is often compared to beauty; it is difficult to define, but you recognize it when you see it. That's why leadership is one of the greatest trusts we can be given. It's rooted in character. Most importantly, it's about how we show up for others every day. Throughout my journey as a pastor, life coach, and Navy chaplain, I have been repeatedly reminded that leadership is proven in small, daily choices. It shows in how faithfully we serve those around us, even when no one is watching, and is built when a leader chooses integrity over convenience, presence over distraction, and service over self-interest. These choices are the ones that define us.

I served with Steve. I am proud to be friends with him, and I appreciate *Helm & Horizon* as a daily guide for Sailors who want to develop as leaders in both skill and character. He has drawn from some of the best leadership thinkers, connected their ideas to strong naval tradition, and presented them in a way that speaks directly to the heart of a leader and to a life in uniform. This book gives leaders the opportunity to pause, reflect, and reset. Each of us needs that rhythm if we are to stay true to our calling.

In the second season of my podcast, *Leadership Spot Check*, we explored fifteen timeless leadership traits that U.S. Marines memorize using the acronym *JJ DID TIE BUCKLE E*: Justice, Judgment, Dependability, Initiative, Decisiveness, Tact, Integrity, Enthusiasm, Bearing, Unselfishness, Courage, Knowledge, Loyalty, Endurance, and Empathy. These traits are more than just words to memorize; they are habits of the heart. They shape our influence, guide how we serve, and determine how we show up every day ready to accomplish the mission. This book mirrors that same spirit. It offers Sailors a chance to reflect and ask tough but essential questions: *Am I leading with integrity? Am I building trust? Am I becoming the kind of leader others can depend on?* The routine of daily reflection here will sharpen your leadership on the deckplates. Still, it will also do something more profound. It will develop consistency, and consistency is the foundation of trust. Leadership rooted in consistency is the kind that lasts well beyond service and extends into every aspect of life: family, community, and vocation. Leadership, after all, isn't just about reaching goals or ticking boxes and it isn' gauged by ribbons on your chest or evaluation bullets on your FITREP. Leadership is about becoming. It's about who you are in the process, not just what you produce at the end.

One of the truths we highlight on *Leadership Spot Check* is that growth never occurs alone. Leadership is not a solo sport. You cannot grow in isolation. We improve best when we reflect together, when we have the courage to ask for feedback, and when we choose to walk the path with others. My encouragement to you is simple: don't rush through this book. Take your time. Let it challenge you, inspire you, and remind you of the leader you are called to be. Share it with your shipmates. Talk about it in the mess decks. Discuss it with your division. Leadership reflection becomes leadership culture when it is shared. Military life is demanding. It tests every part of who we are, physically, mentally, emotionally, and spiritually. It requires us to carry burdens others never see and make decisions others will never understand. But it also gives us the chance to become something greater than ourselves: leaders who stand watch, who bear the weight, who guide others through storms. That's why resources like *Helm & Horizon* matter. They equip us with tools not just to survive but to grow.

As I share these words, I do so with gratitude. Gratitude for sailors who keep stepping forward, day after day, to lead with courage. Gratitude for resources like this one that sharpen our edge. And gratitude for the reminder that leadership is never just about us. It's about those we serve. I commend this book to you. Let it shape you. Let it steady you. Let it remind you that leadership is a sacred trust, one of the greatest we will ever carry. May it help you become the kind of leader who can be counted on in the dark, who embodies integrity when tested, and who leads with strength and humility on every sea.

Helm & Horizon is a gift. I hope that it leaves a lasting impact not only on Sailors but also on the Navy as a whole. May it help you keep a steady hand on the helm and your eyes focused on the horizon.

Dr. Ryan Daffron
U.S. Navy Chaplain & Creator of *Leadership Spot Check*

"We are what we repeatedly do. Excellence, then, is not an act, but a habit."

— Aristotle, *Nicomachean Ethics*

INTRODUCTION

Leadership is a mindset, not a title and not a rank. It is forged in quiet moments on the bridge and sharp ones in combat, tested in the heat of engine rooms, refined in habit on the deckplates. And when the test arrives, it often does so without warning. On the night of June 17, 2017, *USS Fitzgerald* collided with the Philippine-flagged container ship *ACX Crystal* off the coast of Japan. In an instant, routine turned to crisis, as Sailors of every rank and role, officer and enlisted, were thrust into darkness, wading through flooding spaces, assisting the wounded, forcing open jammed hatches, and guiding shipmates to safety with nothing but instinct, training, and resolve. There was no time to look up a manual. They didn't have time to consult doctrine or flip to a chapter on crisis management. What mattered to them was what they had already learned: how to stay calm, act fast, and trust the person beside them. Some led damage control efforts. Some, despite their wounds, guided others through the dark to safety. Some even stayed back in flooded compartments knowingly risking their lives to ensure no one was left behind. These Sailors acted on real, immediate instincts shaped by urgency and tested in motion. It was leadership, plain and raw, born in the moment and proven by action. What those Sailors learned was that leadership doesn't announce itself. It simply shows up. It arrives in the dark, in the flood, in the fight. And when it does, much like the Sailors aboard the *Fighting Fitz*, you'll want to be ready.

My goal with *Helm & Horizon* is simple: take the wisdom of great leaders and make it useful. Right here, right now, for every motivated Sailor who wants to lead with purpose today, grow into a stronger leader tomorrow, and leave a legacy that lasts beyond their time in service. Think of this book as a whetstone. Your opportunity to sharpen your saw. A practical, accessible tool to help you grow, strengthen your habits, and prepare you for the challenges you'll face both on and off duty. Each page is a single entry which pulls from the best minds in war, work, and life boiled down into lessons that speak the language of the Fleet. This curated collection of principles was hand-picked from the most impactful leadership books that span warfighting, mentorship, strategy, resilience, communication, and character, written by some of the most experienced minds across military service, corporate success, sports leadership, psychology, and classical philosophy. Principles tested in command, in combat, in coaching, and in life. And each entry pairs the principle with a real story, some drawn from our Navy's rich history or definitive moments from those who've worn the uniform. Some dive into civilian ideas which are translated into shipboard terms: into quals, divisions, command climate, and the thousand small choices that shape a Sailor's tour. At the bottom of each page, you'll find two ways to act and one final thought to carry forward. Read it before quarters. Add it to the Plan of the Day. Use it to start a Sailor 360 session. Time is tight in the fleet and leadership training can fall behind the pace of the mission. That's why each entry is short. One page. One principle. Something worth using. No fluff. Just the bottom line. Read one a day. And over the next year, your leadership will deepen as your understanding expands.

Many of the principles I have outlined are deliberately revisited throughout the year by design. James Clear, in *Atomic Habits*, reminds us that success doesn't come from a single breakthrough, but from the compounding effect of small, consistent actions over time. This book functions as a daily system meant to provide an anchor for reflection, recalibration, and action. A pause to think, adjust, and lead better. It may not feel like much at first, but over time, it sharpens your thinking, builds your habits, and sets a tone your Sailors will follow. Growth doesn't happen in a rush. Let each entry sink in. Reflect on it. Apply it. Return to it. Admiral James Stavridis, once Supreme Allied Commander of NATO, believes good leaders keep learning and that reading is a duty. He's said it plainly: "Leaders who do not read will eventually be led by those who do." That's the spirit behind *Helm & Horizon*. This book is your leadership library, trimmed to fit the day. In a fast world, where the demands keep climbing, those who lead well are those who never stop learning. The Navy's frameworks give you structure. This book gives you spark. Where those frameworks provide institutional guidance, *Helm & Horizon* provides the daily fuel to put those lessons into action, from the deckplates to the bridge.

Lead with purpose. Take the helm. Keep your eyes on the horizon.

Now let's get to work.

Each of the daily entries follows the same format:

1. **Principle of the Day** – A leadership principle drawn from one of the top leadership texts.

2. **Naval Application** – An interpretation that provides additional thought, insight, and real-world historical or modern examples from the fleet.

3. **Take the Helm** – Two specific applications to lead with this principle in daily naval life.

4. **Eyes on the Horizon** – A closing thought to reinforce the lesson and look forward.

Helm & Horizon is divided into four parts, each building upon the last to create a complete leadership progression. Every chapter is structured to help Sailors grow in self-awareness, strengthen influence, and become the kind of leader others remember long after the tour is over.

PART I: FOUNDATIONS OF LEADERSHIP

Before you can lead others, you must lead yourself. These entries are about owning your standards and demonstrating through daily action that you can be trusted. Because in the Navy, leadership is never just about what you say, but rather what you do when no one's watching.

Chapter 1: Leadership Begins With You

The first thirty days begin with the most important truth in leadership: it starts with you. These entries explore ownership, accountability, discipline, adaptability, and the power of setting the example. From the smallest action to the most visible decisions, you are always leading.

Chapter 2: The Trusted Leader

Trust is the currency of leadership, and it is earned through consistent conduct, sound decision-making, and moral integrity. This chapter dives into the daily habits that build credibility and explores how ethical leadership shapes teams, even in crisis or uncertainty.

Chapter 3: Leadership as a Profession

Leadership is a craft. This chapter focuses on mentorship, lifelong learning, and disciplined excellence. It reflects the ethos that leaders are always in training and that professional growth is inseparable from team success.

Chapter 4: Your Moral Compass

Leaders face hard choices, and this chapter emphasizes the courage to choose what is right over what is easy. Through stories of integrity under pressure and decisions made in the dark, it encourages leaders to hold fast to principle, even at cost.

PART II: LEADING THE TEAM

Once you've established credibility within yourself, your leadership expands outward. This section covers how to build teams that perform even when the leader isn't in the room and how to shape climates where Sailors are empowered to act, speak up, and take pride in the mission.

Chapter 5: Building a Winning Team

No one leads alone. This section offers strategies for developing others, building cohesion, establishing expectations, and fostering a team culture that endures even in the absence of direct supervision.

Chapter 6: Mastering Your Influence

Every leader casts a shadow. This chapter focuses on tone, presence, and the way verbal, nonverbal, and behavioral communication shapes the command climate. Leadership is influence, and influence begins with self-awareness.

Chapter 7: Decisions Under Pressure

This chapter brings the leader into the fog of war. Whether on the bridge, in CIC, or in the heat of a personnel crisis, leaders must make decisions with imperfect information and high consequences. Here, we explore the habits and mindsets that prepare leaders to act decisively.

Chapter 8: Leading Through Change

Change is inevitable. Those who lead through it must model resilience and clarity. Whether facing technological disruption, institutional transformation, or personal adversity, this chapter offers tools to guide teams forward with cohesion.

PART III: THE ENDURING LEADER

Leadership extends beyond your tour and continues in the systems you build, the people you mentor, and the standard you leave behind. This section challenges you to think beyond today's watchbill or command schedule. It focuses on legacy, vision, and strategic thinking preparing you to lead not just through task execution, but through institutional impact.

Chapter 9: The Warfighter's Mindset

Leadership in the Navy is not abstract. It is tied to mission readiness and warfighting excellence. This chapter reinforces the link between character and combat effectiveness, cultivating the mindset of a leader prepared for conflict and peace alike.

Chapter 10: Tactical And Strategic Thinking

The best leaders operate in two modes: managing the moment and preparing for what comes next. Here, we examine how tactical precision and strategic vision must co-exist in decision-making at every level.

Chapter 11: Leadership Beyond the Lifelines

This chapter explores leadership at the enterprise level, where decisions influence policy, partnerships, and the long-term direction of the Navy. It emphasizes the responsibility to think and act institutionally and lead in a way that serves both the Fleet and the broader force.

Chapter 12: Establishing Your Legacy

Leadership does not end when the tour is over. These final entries focus on legacy: mentoring the next generation, living with integrity, and leaving something better behind. This is the capstone of sustained, purpose-driven leadership.

PART IV: CHARTING THE COURSE

This final section will synthesize the lessons of the previous chapters into actionable focus areas for long-term leadership effectiveness and personal development. It is both a reflection and a projection, a reminder that leadership is a continuous journey.

PART I:

FOUNDATIONS OF LEADERSHIP

CHAPTER 1:

LEADERSHIP BEGINS WITH YOU

Day 1 – Leadership is a Mindset, Not a Rank

Turn the Ship Around! – L. David Marquet

"Leadership is not for the select few at the top. In highly effective organizations, there are leaders at every level."

Leadership is not something bestowed by rank or position. It is a mindset and a deliberate choice to take ownership. The traditional military model often places leadership solely in the hands of those at the top, but L. David Marquet identified the flaw in that approach. He realized that in high performing commands, leadership must exist at every level, not just the wardroom. When Marquet assumed command of *USS Santa Fe*, he shifted from a "Leader-Follower" to a "Leader-Leader" model, empowering Sailors to say, "I intend to…" instead of "Request permission to…" That shift built confidence, trust, and transformed *USS Santa Fe* into one of the Navy's best performing submarines. When individuals wait for orders or permission to lead, initiative stalls and problems grow unchecked. But when Sailors are encouraged to take responsibility, make decisions, and act with intent, the command becomes more agile and effective. Marquet's principle is clear: the best organizations are built on shared leadership where accountability is pushed down the chain and not hoarded at the top.

Take the Helm
- **Lead From Your First Watch** – Leadership begins with how you carry yourself every day. Whether on the bridge, in the engine room, or at quarters, step up and take on responsibility.

- **Act Like a Leader Before You Are One** – Take initiative in small ways. Mentor junior Sailors, own your mistakes, and be the one who finds solutions instead of problems.

Eyes on the Horizon
Leadership doesn't come with rank. Initiative, accountability, and the boldness to act is what will distinguish you as a leader. Whether you're a Seaman or a Commanding Officer, leadership is a choice you make every day.

Day 2 – Taking Extreme Ownership

Extreme Ownership – Jocko Willink & Leif Babin

"The leader must own everything in his or her world. There is no one else to blame."

Great leaders take full responsibility for their decisions, their people, and their outcomes. Excuses may feel justified, but they weaken trust and diminish influence. When things go wrong, strong leaders don't deflect or blame, they assess, adapt, and lead forward. This mindset of "extreme ownership" means taking absolute responsibility for everything in your world, regardless of fault. It goes beyond typical accountability by removing blame and excuses entirely. Ultimately, it builds a culture where teams feel supported and accountable, knowing their leader owns both success and failure. Jocko Willink, a Navy SEAL Officer and Task Unit Commander in Ramadi, Iraq, shared a defining moment that captured this principle. After a chaotic clearance operation resulted in a friendly fire incident, Willink was tasked with writing the investigation. Multiple factors contributed: miscommunication, poor coordination, and unclear positioning. But when he briefed his chain of command, Willink didn't spread blame, he stood up and said, "There is only one person to blame for this... me." His ownership earned deep respect and allowed him to implement fixes that prevented future failures.

Take the Helm

- **Own Your Watch, Your Gear, Your Team** – Whether you're standing watch, maintaining equipment, or leading a work center, take full responsibility for everything in your sphere of influence.

- **Take Charge of Your Training** – Instead of waiting for someone to tell you what to learn, actively seek out knowledge, ask questions, and master your craft. Leaders don't wait to be taught. They take ownership of their own development.

Eyes on the Horizon

Excuses erode leadership. Ownership builds it. When you take full responsibility, you gain trust, respect, and control over outcomes.

Day 3 – The Power of First Impressions

The 5 Levels of Leadership – John C. Maxwell

"People naturally follow leaders stronger than themselves. Leadership begins with influence, and influence is earned through respect and trust."

As John C. Maxwell teaches, leadership starts with influence. In fact, leadership begins before you speak. First impressions matter, from the way you carry yourself to how you communicate, observe, and respond in those critical first moments. When then–Commander Michelle Howard took command of *USS Rushmore* in 1999, she was breaking barriers as one of the first Black women to lead a U.S. Navy warship. She wasn't there to make history. She was there to perform. From day one, Howard emphasized readiness, clarity, and excellence. She walked the decks, listened to her crew, and backed her words with action. Her presence was steady, professional, and confident. That first impression became the foundation of a leadership legacy that would carry her to history-making heights, eventually becoming the first female four-star admiral in the U.S. Navy. Her story proves that trust begins the moment you take command, and leadership grows from that first act of credibility.

Take the Helm
- **Start Strong** – Your first words, actions, and appearance set the tone. Walk in prepared, engaged, and professional from day one. People follow those they believe in, and belief starts with what they see in you at the outset.

- **Be Dependable** – Leaders who deliver consistently earn trust fast. Show up on time, ready to execute, and always squared away.

Eyes on the Horizon
Your reputation begins the moment you step into the arena. Lead with integrity from day one and trust will build the credibility that carries you forward.

Day 4 – Self-Discipline as Leadership

Can't Hurt Me – David Goggins

"You are in danger of living a life so comfortable and soft, that you will die without ever realizing your true potential."

Self-discipline is the bedrock of leadership. By showing up prepared, controlling your emotions, and upholding high standards, you as a leader earn respect without ever having to demand it. Without discipline, talent fades, authority weakens, and potential goes unrealized. In the Navy's most grueling training pipeline, Basic Underwater Demolition/SEAL (BUD/S) training, this principle is tested in the harshest conditions imaginable. Candidates are pushed to their physical and mental limits through cold, exhaustion, and stress. What determines who makes it through? Not size or speed, but the daily decision to stay focused, composed, and mission ready. That is self-discipline in action. David Goggins' experience as a SEAL leader drives this point home. He watched firsthand how discipline, not motivation, separated future leaders from those who rang the bell. What instructors valued wasn't who was loudest or strongest, but who could hold the line under pain, pressure, and fatigue. Those are the leaders who can be trusted in combat, and in command. Discipline is what gives leaders the freedom to act with clarity, to earn trust, and to rise when others fall apart.

Take the Helm
- **Stay Ready for the Fight** – A disciplined Sailor keeps qualifications current, gear in check, and knowledge sharp to be mission-ready at all times

- **Set the Standard** – If you expect on-time watch reliefs, squared-away uniforms, and professional conduct, then uphold these standards yourself.

Eyes on the Horizon
Discipline isn't restrictive, but rather it's the path to control. When you master yourself through discipline, you earn the clarity and consistency needed to lead others with purpose.

Day 5 – Leadership Through Actions

Leaders Eat Last – Simon Sinek

"The true price of leadership is the willingness to place the needs of others above your own."

The heart of leadership is found in how you serve, how you show up, and how you lead by example. Simon Sinek argues that the most effective leaders prioritize their people's well-being and set the tone by how they show up every day. That principle was powerfully demonstrated by Admiral Chester Nimitz after Pearl Harbor. Rather than blame others for the Navy's unpreparedness, Nimitz placed the needs of his Sailors above his own reputation or comfort. He showed up, listened, and immediately began steering the fleet toward recovery. He walked the piers, stood on battered decks, and made his presence felt. His steady leadership gave Sailors a reason to believe they could still win. Just as Sinek describes, Nimitz earned trust not with speeches, but with visible commitment and composed action. By showing up and refusing to flinch in crisis, he re-established confidence across the fleet. He laid out the emotional and operational foundation that led to a successful Pacific campaign, proving that when leaders act first, others rise with them.

Take the Helm
- **Your Division Reflects You** – If you're squared away, your team will be too. If you cut corners, so will they. Set the example in uniform, work ethic, and professionalism because your Sailors will follow what you do, not what you say.

- **Stay Calm, Stay Ready** – A leader's composure in casualties, high-stress evolutions, and day-to-day operations sets the tone. Whether you're taking the conn or handling a maintenance issue, lead with confidence and control.

Eyes on the Horizon
What you model through action teaches more than any directive. Your actions set the standard long before your words ever reach the deckplates.

Day 6 – Doing the Right Thing When No One is Watching

The 7 Habits of Highly Effective People – Stephen R. Covey

"Integrity is what we do, what we say, and what we say we do."

Private choices often reveal more about a leader than public gestures. The decisions you make when no one is watching reflect who you truly are and what kind of leader you will become. Trust isn't earned by words alone, but by the quiet consistency of doing things right, day in and day out. Those who cut corners may find short-term ease, but they lose something far more valuable: credibility. This truth is lived daily aboard our submarines, where Sailors operate under the surface, out of sight, and often under immense pressure. Every procedure, log entry, and equipment check must be carried out with precision, not because someone is watching, but because lives depend on it. There's no room for shortcuts in the silent service. The nuclear submarine community exemplifies accountability in its purest form: a culture where each Sailor owns their actions, not for praise, but because it's the standard. That discipline in isolation is one of the clearest demonstrations of trust, professionalism, and character in our Navy today.

Take the Helm
- **Prove Self-Discipline When No One is Watching** – Whether performing maintenance, standing watch, or reporting issues, take ownership even when no one is watching. Respect is built when Sailors know they can rely on you even when no one else is around.

- **Hold the Line on Standards** – Integrity means ensuring safety, accuracy, and readiness, even when there's an easier path.

Eyes on the Horizon
Character is forged in solitude long before it's seen in action. The right decision is the one you'd make even if no one ever found out.

Day 7 – The Mission Before Self

The Mission, the Men, and Me – Pete Blaber

"It's not about you. It's about the mission, and it's about the team. When leaders understand this, everything else falls into place."

The strongest leaders elevate others and drive the mission forward without seeking the spotlight. They don't chase credit or protect pride when responsibility is on the line. Instead, they bring focus, humility, and discipline to the moment, making decisions that serve a higher purpose. Leadership grows when others know you're committed to something greater than yourself. Admiral Raymond Spruance embodied this mindset during one of the most pivotal moments in naval history: the Battle of Midway. Days before the operation, he was unexpectedly assigned to command Task Force 16 after Admiral William Halsey fell ill. With limited carrier experience and enormous stakes, he could have hesitated, but he didn't. Spruance trusted his team, made bold decisions under pressure, and kept the mission at the center of every action. His calm leadership and early strike orders led to a decisive victory that turned the tide of the Pacific War. Spruance didn't seek glory. Results mattered more than recognition.

Take the Helm

- **Put the Team First** – Leaders eat last, stay late, and take care of their people before themselves. If the team succeeds, you succeed.

- **Make Decisions for the Mission, Not Yourself** – Whether standing watch or leading a division, prioritize what's best for the ship and crew over what's easy or comfortable.

Eyes on the Horizon

Real leadership begins when personal gain takes a back seat to the needs of the team. The best leaders make sacrifices so their teams can succeed.

Day 8 – Resilience in Leadership

The Obstacle Is the Way – Ryan Holiday

"The impediment to action advances action. What stands in the way becomes the way."

Anyone can lead in calm conditions. Real leadership is tested when the pressure mounts. In the Navy's high-pressure world of deployments and crises, setbacks are inevitable. What separates great leaders is resilience. It's the ability to face hardship head-on, adapt to challenges, and keep moving forward. Ryan Holiday teaches that obstacles don't block the path; they *are* the path. Leaders must learn to treat adversity as the proving ground for growth and strength. Few leaders embodied this more powerfully than Admiral James Stockdale. As a Prisoner of War (POW) in Vietnam for over seven years, Stockdale endured torture, isolation, and uncertainty. Yet he refused to be broken. Despite his situation, he established a code of conduct for fellow prisoners, resisted enemy exploitation, and became a moral center in the face of chaos. What set Stockdale apart was not just survival, but the clarity and conviction with which he led through darkness. His example reminds every Sailor that leadership under pressure means more than staying calm. It means turning pain into purpose and hardship into strength.

Take the Helm
- **Set the Tone During Casualties and Emergencies** – When facing flooding, fire, man overboard, loss of steering, or other tragedies, your ability to remain calm and issue clear, confident orders will stabilize your team and save lives.

- **Debrief and Refit After Every Setback** – After failed inspections, drills, or evolutions, conduct an honest debrief, identify gaps, and implement corrective action. Growth comes from how your team responds to adversity.

Eyes on the Horizon
Comfort reveals little. Crisis reveals everything. Lead with calm, recover with purpose, and build a crew that's stronger on the other side.

Day 9 – Humility as a Strength

Team of Teams – General Stanley McChrystal

"It takes a lot more courage to admit you don't know something than to bluff your way through."

General Stanley McChrystal highlights how the most effective leaders in modern combat environments weren't those who had all the answers, but those who were honest enough to admit when they didn't and humble enough to learn fast. He recounts leading the Joint Special Operations Task Force in Iraq, where rigid, top-down leadership structures were failing against the speed and unpredictability of enemy networks. McChrystal realized that his own effectiveness hinged on humility, accepting input, empowering others, and creating a culture where learning beat ego. When leaders think they already have the answers, they may stop listening and once they stop listening, they stop learning. When you choose humility over image, you unlock better decision-making and build stronger teams. A leader who listens, who credits others, and who isn't afraid to ask questions builds credibility that lasts. In fast-paced, high-stakes environments, no one has time for posturing. The leaders who grow, who help others grow, are those who admit what they don't know and lead by example in learning. McChrystal's experience reminds us that humility is strategic. It allows teams to adapt, innovate, and succeed together. And in the Navy, that's the kind of leadership Sailors never forget.

Take the Helm
- **Listen More, Talk Less** – A humble leader absorbs knowledge from Chiefs, junior Sailors, and peers rather than assuming they know everything. Keep quiet confidence in your tactical toolbelt.

- **Credit the Team, Not Yourself** – When things go right, recognize your Sailors' efforts. When things go wrong, own the mistakes and learn.

Eyes on the Horizon
Ego clouds a leader's vision. Humility brings clarity. The best leaders never stop learning.

Day 10 – Decisiveness in the Face of Uncertainty

The 33 Strategies of War – Robert Greene

"Uncertainty is a permanent part of war. You must learn to act despite it."

Leadership in naval operations demands action often before perfect clarity arrives. Robert Greene reminds us that indecision can be just as dangerous as the wrong decision, especially when time is critical. In 2006, the crew of *USS Winston S. Churchill* faced this truth firsthand when they identified a suspicious vessel off the coast of Somalia. With limited intelligence, high risk, and no guaranteed outcome, the commanding officer had to choose between delay and action. He didn't wait. He trusted his training, assessed the situation, and moved decisively. That willingness to act under pressure embodies the core of leadership: making the best call possible, even when the outcome is unclear. The boarding was executed without casualties, the threat was neutralized, and the crew's confidence in their leadership was reinforced. *USS Churchill's* actions show how decisive leadership turns risk into opportunity, builds trust through clarity, and prevents hesitation from becoming failure. Leaders must remember: no environment will ever be fully certain. But when Sailors see their leaders act with calm, deliberate purpose, even in uncertainty, they will follow.

Take the Helm

- **Assess, Decide, and Own It** – Perfect information doesn't exist. Make the best call based on what you know and take responsibility for the outcome.

- **Train for Quick, Clear Decisions** – Whether taking the conn, leading a damage control effort, or handling a personnel issue, train your instincts to make timely, effective choices.

Eyes on the Horizon

Leaders don't wait for certainty. Analyze, decide, act. When others freeze, your resolve can move the mission forward.

Day 11 – Confidence Without Arrogance

With the Old Breed – Eugene B. Sledge

"We had to learn quickly that courage wasn't bravado. It was doing your job despite the fear."

Eugene B. Sledge reflects on the brutal lessons of war in the Pacific where leadership was tested, not by bravado, but by steady courage. He describes how Marines respected leaders who stayed composed under fire and who carried the weight of responsibility without making it about themselves. This balance is crucial in the demanding environment of naval operations. A Sailor leading a damage control team during a fire must project confidence to reassure their team and direct their efforts effectively; however, they must also be humble enough to listen to the input of experienced Sailors who may have valuable insights into the situation. Overconfidence could lead to overlooking critical details or dismissing valuable suggestions, potentially exacerbating the crisis. The effective naval leader understands that true strength lies in balancing decisive action with the humility to recognize the value of every team member's contribution, especially in high-stakes situations where lives and the ship's safety are on the line. As Sledge's account reminds us, the leaders who earn the most trust are those who show their courage not in volume, but in calm, competent command even when the stakes couldn't be higher.

Take the Helm
- **Project Command Presence** – Speak with clarity and authority but never dismiss the input of others. Your Sailors will follow a leader who is steady, not boastful.

- **Seek Advice, Then Decide** – Confidence means trusting yourself to make the call, but arrogance ignores valuable insights. A great leader gathers knowledge before acting.

Eyes on the Horizon
Confidence commands respect and arrogance destroys it. Stay steady, listen well, and lead decisively.

Day 12 – The Power of Adaptability

Mindset – Carol S. Dweck

"Becoming is better than being."

Carol S. Dweck's research on the difference between fixed and growth mindsets reveal a truth every leader must embrace: your potential isn't predetermined. How you respond to challenges shapes what you become. Leaders who resist change stagnate, but leaders who welcome uncertainty as a chance to evolve not only grow themselves but move their teams forward. The Navy demonstrated this mindset during one of its most defining transformations. In the early stages of World War II, naval doctrine still favored battleships as the backbone of sea power. But after Pearl Harbor shattered that belief, Navy leadership didn't cling to the past. Leaders adjusted course and adapted to meet new threats. Rather than double down on outdated models, the Navy shifted its focus to aircraft carriers and airpower, radically changing how it projected force. This wasn't a smooth or easy pivot, but it was necessary. Embracing a growth mindset meant rethinking strategy, retraining crews, and adjusting operational priorities on the fly. That decision not only turned the tide of the war in the Pacific but also reshaped the future of naval warfare. Adaptability didn't come from knowing all the answers, it came from a willingness to learn fast and lead through change. That's Dweck's lesson in action: success favors those who are becoming, not those who believe they've arrived.

Take the Helm
- **Adopt a Growth Mindset in Your Watchstanding and Training** – Be open to feedback, new procedures, and evolving tactics. Flexibility leads to mastery.

- **Lead Change from the Front** – When policies, platforms, or missions change, don't resist. Help your team adjust, implement, and own the new direction.

Eyes on the Horizon
The Navy doesn't stand still and neither should you. Adaptability ensures you remain mission-ready in every evolution, deployment, and command.

Day 13 – Emotional Intelligence in Leadership

The Obstacle Is the Way – Ryan Holiday

"You don't control the situation. You control how you respond to it."

Emotional intelligence is one of the most powerful and often overlooked tools in a leader's arsenal. In high-stakes environments, leaders don't always control the circumstances, but they control their presence. As Ryan Holiday points out, calm in the storm starts with self-mastery. MCPON Mike Stevens believed exactly that. As the 13th Master Chief Petty Officer of the Navy, he often said, "Before you can lead people, you have to care about them." His development of the CPO 365 program, the foundation of today's Sailor 360, placed deliberate focus on emotional intelligence as a core component of deckplate leadership. It taught Chiefs not just to enforce standards, but to listen with intention, recognize distress, and lead with presence. Whether it's a struggling Sailor, frustrated peer, or Chief overwhelmed by new responsibility, Stevens pushed leaders to look beyond performance alone and understand the person behind it. That's emotional intelligence in action: knowing when to challenge, when to support, and how to manage your own tone and reaction in the process. Emotional intelligence doesn't weaken your leadership. Trust grows when leaders understand and manage emotions well

Take the Helm
- **Master Your Emotions in High-Stakes Moments** – Read the room. Whether during a complex evolution or casualty response, calm leaders stabilize the team and maintain control of the situation.

- **Adjust Your Approach Based on the Sailor and the Situation** – Firmness earns respect, but empathy builds loyalty. Know when to coach, when to correct, and when to listen.

Eyes on the Horizon
In war, operations, and shipboard life, you won't always control the circumstances, but you can always control yourself. And in that control, your team finds strength.

Day 14 – The Value of Mentorship

The 5 Levels of Leadership – John C. Maxwell

"The highest level of leadership is developing leaders who develop other leaders."

John C. Maxwell teaches that the most effective leaders multiply themselves by developing others, creating a legacy that endures beyond any single billet or command tour. Admiral Chester Nimitz lived this principle. As Commander of the Pacific Fleet in World War II, he didn't simply lead from the top, he built a generation of leaders who would carry the war forward. He invested in officers like Admiral Raymond Spruance, the calm and analytical architect of Midway; Admiral William "Bull" Halsey, the aggressive and bold commander who energized fleet morale; and Vice Admiral Charles Lockwood, who revitalized the submarine force into a strategic weapon. Nimitz didn't choose them because they were fully polished, but because he recognized their potential and made it his mission to cultivate it. Nimitz's mentorship created a ripple effect of confident, competent decision-makers who shaped the course of naval operations. The same principle holds true at every level of the fleet today. Commands that prioritize mentorship are stronger, more resilient, and better prepared for the future. Sailors step up when they know someone believes in their growth. Victory in the Pacific was only part of what Nimitz achieved. His greatest impact came from empowering others to lead boldly in his wake.

Take the Helm

- **Mentor Through Delegation and Feedback** – Give your Sailors opportunities to lead and grow, then guide them with direct, honest feedback.

- **Pursue Your Own Mentorship** – Find a Chief, Officer, or Senior Sailor who challenges you and helps refine your leadership.

Eyes on the Horizon

Great leaders do more than stand the watch. They prepare the next generation to take the helm. Mentorship is leadership that endures.

Day 15 – Small Acts, Big Impact

Make Your Bed – Admiral William H. McRaven

"If you want to change the world, start by making your bed."

Admiral William H. McRaven reminds us that success in the big things starts with consistency in the small ones. That principle shows up in every space, every checklist, every watch turnover. The goal isn't robotic compliance. Deliberate habits prepare you to perform at your best. The Navy's inspection processes, from engineering to combat systems to supply, reflect this truth. Commands that treat discipline as a daily standard, not a last-minute scramble, earn trust, perform under pressure, and remain ready when it counts. A command that embraces disciplined routines doesn't do it for inspection day; they do it because it defines who they are every day. Tight logs, clean workspaces, detailed tag-outs, and squared-away divisions speaks volumes about the state of your leadership. Crews that internalize high standards develop pride, precision, and resilience. Just like McRaven's bed-making advice, small actions compound into larger excellence. Naval operations demand precision, and the most disciplined leaders meet that demand by embodying the standard they uphold.

Take the Helm
- **Discipline Starts With You** – Master the fundamentals: uniform standards, accountability, and personal readiness to set the tone for your team.

- **Build a Culture of Excellence** – Demand discipline in maintenance, watchstanding, and procedures because "good enough" is never good enough.

Eyes on the Horizon
Lasting discipline is built on daily habits that don't waver. Lead by embracing standards and your team will rise to meet them.

Day 16 – The Power of Clear Communication

Crucial Conversations – Kerry Patterson, Joseph Grenny, Ron McMillan, Al Switzler

"The measure of success is not whether you have tough conversations, but whether you handle them well."

Leadership cannot exist without clear communication. Success doesn't hinge on whether conversations happen, but rather on how well they're executed. The 2017 collisions of *USS Fitzgerald* and *USS John S. McCain* are tragic reminders of what happens when communication breaks down. In both cases, poor coordination and lack of challenge up and down the chain led to catastrophic outcomes. These incidents point not just to seamanship errors, but to failures in communication under pressure. Watchstanders didn't speak up. Orders weren't confirmed. Control of critical systems, like steering, were misunderstood or never clearly transferred. These collisions claimed the lives of 17 Sailors and exposed how quickly uncertainty turns into tragedy when leaders don't foster a culture of clear, direct, closed-loop communication. In contrast, commands that prioritize precise orders, repeat-backs, and shared understanding perform with discipline and confidence, even under pressure. Whether issuing commands on the bridge, relaying damage control reports, or debriefing a mission, your words matter. In critical moments, when it matters most, clarity is a duty every leader must uphold

Take the Helm
- **Communicate with Purpose** – Whether issuing a maneuvering command from the bridge or briefing a fire party, use standardized, professional language to eliminate ambiguity and drive action.

- **Confirm and Repeat Critical Instructions** – In combat, casualty control, or restricted maneuvering scenarios, require repeat backs or acknowledgments to ensure orders are received and understood before execution.

Eyes on the Horizon
Leadership strength is revealed in the clarity of your message under pressure. Say what you mean. Verify it was received. Watch your team execute with confidence.

Day 17 – Leading Through Crisis

Leadership in War – Andrew Roberts

"The moment of crisis does not create character. It reveals it."

Crisis strikes without warning, breaking routine and testing the core of leadership. Andrew Roberts teaches that moments of crisis reveal leadership rather than create it. In 1967, a catastrophic fire aboard *USS Forrestal* began with a rocket misfire, setting off a chain of explosions across the flight deck. Amid flames, disorientation, and live ordnance cooking off, Sailors didn't panic. Calm leadership guided their actions with steady hands. Officers and junior crew responded with focus and grit, proving that real leadership is revealed when preparation meets pressure. *USS Forrestal* became a turning point not just for damage control doctrine, but for understanding what leadership under fire looks like. The tragedy led to sweeping changes in firefighting procedures and training, most notably the creation of the Navy's Afloat Firefighting School and the standardization of damage control qualifications fleetwide. But what emerged wasn't just better equipment, but a new culture of preparedness driven by leaders. In any crisis, whether on the flight deck or during combat, leaders who stay composed and act with purpose give their crews something to anchor to. The fire revealed the truth: leadership isn't tested by calm seas, but by fire.

Take the Helm
- **Rehearse Crisis Response Until It's Muscle Memory** – Drill general quarters, damage control, and man overboard procedures relentlessly. When crisis hits, training takes over.

- **Command the Atmosphere with Calm Authority** – In every emergency, how you issue orders, control tone, and set the pace will either steady your crew or shake their confidence.

Eyes on the Horizon
No crisis asks for an appointment. They strike fast and test everything you've built. When the alarm sounds, your crew will look to you. Be ready to lead with discipline, composure, and clarity.

Day 18 – Leading by Serving Others

Leading With the Heart – Mike Krzyzewski

"A leader is someone who puts the needs of his team above his own. That's the only way a team can grow."

True leadership is rooted in service. Coach K recounts the 1994 Duke basketball season when he stepped away from coaching due to serious health issues. He had the option to push through and continue, but he realized that doing so would have put his players, staff, and the program at risk. His decision wasn't about preserving his pride but about protecting the team's development and future. That kind of selfless judgment is the essence of servant leadership. This shows up when a leader steps back to let a junior Sailor lead a brief, mentors someone through tough feedback instead of pushing them aside or sacrifices personal time to ensure their division is set up for success. Like Coach K, naval leaders must sometimes forgo control or credit to grow the next generation. Leaders who prioritize their Sailors' development, welfare, and readiness over their own convenience build unshakable trust. And that trust turns into initiative and enduring excellence. That's how serving others becomes your most powerful form of leadership.

Take the Helm
- **Prioritize Sailor Readiness Over Personal Comfort** – Whether distributing gear, qualifying personnel, or adjusting watch rotations, ensure your team is set before tending to your own needs.

- **Model Selfless Service on Deckplates and in the Field** – Step into the difficult jobs whether it's standing the extra watch, staying late for maintenance, or supporting a Sailor in crisis. Actions define servant leadership.

Eyes on the Horizon
True leadership isn't found behind a desk. Real leadership takes shape where sweat and grit meet, standing shoulder to shoulder with your Sailors. When your crew sees you serving them, not yourself, they'll follow you anywhere. Serve first, lead always.

Day 19 – Accountability in Leadership

Resilience – Eric Greitens

"You will make mistakes. The question is: will you own them, learn from them, and grow stronger because of them?"

Eric Greitens reminds us that there's no room for blame-shifting at the top. Leaders must own every outcome in their sphere, good or bad. In 2001, Commander Scott Waddle of *USS Greeneville* learned this at the highest cost. During a maneuver to demonstrate submarine capabilities, *Greeneville* accidentally struck and sank the Japanese fishing vessel *Ehime Maru*, killing nine civilians. In the aftermath, Waddle did not hide behind process, staff error, or political pressure. He took full responsibility publicly, directly, and without excuse. That decision didn't protect his career, but it preserved something far more important: his integrity. It's easy to acknowledge mistakes. Facing the consequences shows true accountability. Waddle expressed deep remorse, engaged with the victims' families, and accepted the weight of command without deflection. Though he was never formally awarded by the Navy for his response, he was later recognized by leadership institutions, including invitations to speak at military academies, ethics conferences, and leadership summits around the world. His actions became a case study in moral leadership. Whether in a small mishap or a high-stakes failure, your team is watching. They don't expect perfection but they expect ownership. Leadership without accountability is hollow. Leadership with it earns lasting respect.

Take the Helm
- **Own Mistakes Publicly, Fix Them Promptly** – If a drill, inspection, or evolution falls short, acknowledge your role, brief the team, and implement corrections. Your crew will respect you more for it.

- **Enforce Standards with Consistency and Purpose** – Whether it's watchstanding, maintenance, or conduct, build a division where accountability is expected, and excuses don't survive muster.

Eyes on the Horizon

The core of accountability is owning the outcome, good or bad. Your Sailors will follow a leader who owns the mission even when it goes off course.

Day 20 – Act in the Face of Ambiguity

The 7 Habits of Highly Effective People – Stephen R. Covey

"Proactive people focus their efforts on what they can do, rather than what they can't."

Stephen R. Covey teaches that effective people focus on what they can control rather than waiting for perfect direction. That principle was on full display in 1958, when *USS Nautilus* undertook the first ever under ice transit of the Arctic. With no template, no precedent, and limited navigational certainty, the mission demanded proactive thinking at every watch station. Sailors couldn't afford to wait for orders. They had to assess, adjust, and solve problems in real time, often in conditions no one had faced before. The success of the *USS Nautilus* mission rested on more than advanced technology. A crew empowered to lead at every level made the difference. Officers and enlisted alike demonstrated initiative, turning an uncharted environment into a breakthrough achievement in naval warfare. The Navy rewards those who lead without being asked, who take action instead of waiting, and who turn ambiguity into opportunity.

Take the Helm

- **Act on Deficiencies Before They Cascade** – If you notice an unreported equipment fault, unsafe practice, or procedural lapse, fix it or elevate it immediately. Your initiative may prevent a mishap, inspection failure, or an unknown threat from becoming a crisis.

- **Plan Two Steps Ahead of the Watchbill** – Whether preparing for a training evolution, a maintenance availability, or an upcoming deployment, anticipate what your team and command will need before the question is asked.

Eyes on the Horizon

Leaders are remembered for solving problems, protecting the mission, and staying ahead of the storm. They act with purpose in the face of uncertainty, knowing that momentum keeps the mission moving forward.

Day 21 – The Power of Delegation

Multipliers – Liz Wiseman

"The best leaders are not geniuses with a thousand helpers. They are genius makers."

Delegation is a leadership skill rooted in trust and the wisdom to empower others. Liz Wiseman explores the idea that true leadership isn't about being the smartest person in the room, but about creating an environment where others rise to their full potential. Delegation, in this view, is unlocking capability. Wiseman draws a clear line between two types of leaders: Diminishers, who believe they must drive every decision, and Multipliers, who amplify the intelligence and capability of those around them. One standout example is Lutz Ziob, former general manager of Microsoft Learning. Instead of issuing directives, Ziob fostered a culture of inquiry. He regularly asked his team: "What do you recommend?" signaling trust while inviting ownership. Over time, this approach empowered his team to take initiative, challenge assumptions, and lead major strategic changes on their own. This empowerment enabled Microsoft Learning to scale more effectively and maintain relevance in a rapidly evolving tech environment. Today, effective delegation in the Navy is essential for continuity, growth, and mission success. When you empower others, you multiply leadership, develop talent, and move from being a doer to being a builder of teams.

Take the Helm
- **Delegate to Develop Bench Strength** – Assign authority during drills, inspections, or planning meetings to help Sailors build leadership confidence and competence.

- **Trust, Then Train** – Let your junior leaders take ownership of key duties, and coach them through successes and failures. That's how leaders are made.

Eyes on the Horizon
Delegation isn't just about getting things done. When done right, delegation ensures that leadership skills grow in those who will eventually take the next watch. Let go to lead forward.

Day 22 – Handling Criticism as a Leader

Daring Greatly – Brené Brown

"If you're not in the arena getting your ass kicked, I'm not interested in your feedback."

Criticism is the tax of leadership. Brené Brown reminds us that anyone who dares to lead boldly will be challenged, second-guessed, and judged. Admiral "Bull" Halsey lived in that arena. His high-risk decisions in World War II often drew sharp criticism. After the deadly losses from Typhoon Cobra in 1944 where three destroyers sank and 790 Sailors were killed, Halsey faced a formal Court of Inquiry. The panel harshly criticized his decision to sail the Third Fleet into the heart of a typhoon, citing failure to exercise due caution. Halsey didn't deflect or blame others. He stood accountable, absorbed the critique, and conducted an honest assessment of his decision-making. While he was not relieved of command, the inquiry shaped how he approached subsequent operations. He continued to lead his fleet through the final campaigns of the Pacific but with a deeper awareness of risk and responsibility. Halsey knew that feedback was part of the job, but retreating in fear of judgment was never an option. Leading well requires the courage to withstand critique and the resilience to keep serving through it.

Take the Helm

- **Filter Feedback Through the Mission Lens** – Evaluate criticism based on its impact on readiness, safety, and team effectiveness. Feedback that sharpens your watch team or evolution is worth keeping.

- **Maintain Composure Under Scrutiny** – Whether you're briefing the CO, receiving a spot check, or being challenged in a debrief, respond with professionalism, not ego.

Eyes on the Horizon

Every leader in uniform will face critique. How you handle it is what your Sailors will remember. Refine what's helpful, ignore the noise, and let your response show you're ready for the next level of leadership.

Day 23 – Leading Through Failure

The Obstacle Is the Way – Ryan Holiday

"The obstacle in the path becomes the path."

Failure often serves as the course correction that shapes better leaders. Ryan Holiday reminds us that setbacks, when met with discipline and perspective, become the very ground leaders must walk to improve. The tragic loss of *USS Thresher* in 1963 tested the Navy in exactly that way. After a catastrophic mechanical failure during deep dive trials claimed the lives of 129 Sailors, the Navy responded not with deflection, but with resolve. Leadership at every level accepted responsibility, studied the failure in depth, and launched the Submarine Safety (SUBSAFE) program: a sweeping overhaul of submarine safety, maintenance, and quality assurance. The SUBSAFE program created a standard of excellence built on the ashes of failure. No SUBSAFE-certified submarine has been lost since, a testament to what happens when leaders choose transparency, learning, and action over blame or denial. Failure doesn't define your legacy. How you respond to it does. Great leaders don't avoid failure. Leading through setbacks makes the ship, the crew, and the institution stronger for the next generation.

Take the Helm
- **Debrief Failures with Purpose** – When an evolution fails, walk through it step-by-step, identify breakdowns, and apply the lessons. No finger-pointing, just forward movement.

- **Model Accountability When Things Go Wrong** – If you were responsible, say so. Then lead the fix. Your Sailors will learn more from your recovery than your perfection.

Eyes on the Horizon
Failure in the fleet is a signal not a sentence. Those who lead through it with humility and purpose become the leaders everyone follows when it matters most.

Day 24 – The Strength of Decentralized Command

Turn the Ship Around! – L. David Marquet

"Great leaders don't give orders. They create environments where their teams think and act like leaders."

The most effective naval leaders cultivate commands where Sailors are empowered to lead at every level. L. David Marquet challenges the traditional top-down approach and shows that real strength lies in pushing authority down the chain. The Navy has proven this across multiple domains and decades. In Operation Desert Storm, carrier air wings operated in high-tempo environments with tactical control delegated to squadron and strike leaders. Clear intent, not constant oversight, allowed for faster decisions and real-time adaptability. In special operations, SEAL platoons execute missions built on decentralized execution, where every operator is trained and trusted to lead when conditions shift. Across every form of naval operations, the pattern holds: when leaders clarify intent and step back, capable teams step up. Today's operational environment demands decentralized command as a core element of strategy. Trust multiplies leadership. It empowers speed, sharpens judgment, and ensures resilience in uncertain environments.

Take the Helm

- **Empower Sailors with Ownership** – Assign watchstations, maintenance responsibilities, or planning roles that allow your team to take initiative and lead within their scope of authority.

- **Frame the Mission with Commander's Intent** – Don't just issue orders, explain the objective and desired end state so your Sailors can make sound decisions when you're not standing next to them.

Eyes on the Horizon

Commands that trust their people move with confidence. Train your Sailors to think, act, and lead because one day, they'll be the ones issuing the orders. When every Sailor leads, the team wins as one.

Day 25 – The Role of Patience in Leadership

The 5 Levels of Leadership – John C. Maxwell

"A leader's growth is a process, not an event."

Lasting leadership comes from deliberate growth over time. John C. Maxwell reminds us that lasting leadership takes time, intentional effort, and consistency. Impatient leaders often expect instant progress, pushing too hard and burning out their teams. Admiral Hyman Rickover, the architect of the nuclear Navy, understood that excellence demands patience. Over decades, he battled resistance, refined systems, and built a nuclear Navy from scratch, not by rushing, but by holding the line on rigorous standards and long-term vision. His approach did not rely on shortcuts. Building a safer, smarter fleet was always the goal. Rickover's leadership style was demanding but deeply committed to the process of development. He personally interviewed officers, scrutinized procedures, and built a culture of accountability through relentless attention to detail. Passive waiting did not define his patience. He stayed disciplined and deliberate, trusting that real progress comes through effort. Because of his approach, the nuclear submarine force remains one of the most advanced and safest arms of the Navy. Great leaders understand that real results take time and the strongest teams are shaped through steady guidance, not rushed expectations.

Take the Helm
- **Invest in Long-Term Readiness** – Train your division thoroughly, qualify your Sailors the right way, and focus on sustainable excellence, not just quick wins for inspections or evals.

- **Show Patience That Guides** – During high-stress evolutions, give clear direction, allow room for correction, and stay steady. Your response sets the tone. Don't rush judgment or crowd the moment.

Eyes on the Horizon
Cultivate patience by listening before reacting, stepping back before stepping in, and remembering that growth takes time. Lead with purpose and you'll build a command climate that outlasts your tour.

Day 26 – Courage in Leadership

Call Sign Chaos – Jim Mattis and Bing West

"You must be willing to make tough decisions and take full responsibility for them. That's what leadership requires."

Jim Mattis and Bing West reminds us that leadership requires the willingness to make hard decisions and accept the weight of their outcomes. In 2020, Captain Brett Crozier of *USS Theodore Roosevelt* faced exactly that kind of test. As COVID-19 spread aboard the carrier, infecting dozens of Sailors, Crozier assessed that the chain of command was moving too slowly to protect his crew. He bypassed traditional protocols and sent a four-page memo pleading for urgent action, which was leaked to the press. His decision conflicted with Navy communication procedures and frustrated senior leaders leading to his removal from command. But Crozier didn't act out of defiance but from duty to protect his crew. He knew the consequences yet prioritized his Sailors. Crozier's decisive action sparked national debate, led to a Navy-wide review of pandemic response readiness, and earned him deep respect across the fleet. His crew gave him a standing ovation as he left the ship, proof of the trust he had built. It's that kind of moral courage that inspires Sailors to trust their leaders when it matters most.

Take the Helm
- **Make the Right Call, Even When It's Hard** – Whether confronting a toxic Sailor or challenging a flawed plan, stand your ground and do what's best for the crew and mission.

- **Model Courage in Routine and Crisis** – Show up in tough moments, back your team when it counts, and lead with integrity whether the stakes are high or routine.

Eyes on the Horizon
Courage often means standing alone for what's right. Lead with conviction, not caution, when the pressure mounts and your Sailors look to you for strength.

Day 27 – The Danger of Complacency

Discipline Equals Freedom – Jocko Willink

"Don't let your guard down. Ever. Stay on the path. Stay disciplined."

Complacency doesn't announce itself. It waits and creeps in through routine, unchecked assumptions, and the false belief that "good enough" is still good. Jocko Willink teaches that discipline is what keeps you sharp, prepared, and ready because the cost of letting up isn't always immediate, but it is always dangerous. That reality was made tragically clear in 2000 when *USS Cole* was attacked while refueling in Yemen. Seventeen Sailors lost their lives and dozens more were wounded in a suicide bombing that exploited gaps in force protection and situational awareness, gaps rooted in routine and overconfidence in a high-threat port. The attack on *USS Cole* became a stark reminder that readiness must never go on autopilot. The Navy responded with sweeping security reforms, but the greater lesson still stands: the enemy watches for complacency. Strong leaders don't wait for crisis to adjust. They uphold discipline in the calm to be ready in the storm. Whether leading a watch team, prepping for inspection, or operating in a war zone, vigilance must be daily because the moment you relax your standards could be the moment you need them most.

Take the Helm
- **Demand Discipline in Every Evolution** – From weapons handling to watchstanding, ensure your team doesn't just "check the box" and that they execute with intent and focus.

- **Reevaluate SOPs and Routines Regularly** – Challenge standard procedures to ensure they still reflect the threat environment and operational needs.

Eyes on the Horizon
In combat, in port, and underway, complacency is the enemy. Discipline is what keeps you sharp and alive.

Day 28 – The Power of Recognition

Leaders Eat Last – Simon Sinek

"The more we give recognition, the more people want to step up and do even more."

Rather than a soft tactic, recognition is a proven way to multiply a team's effectiveness. Simon Sinek reminds us that when leaders consistently acknowledge effort, progress, and excellence, it energizes teams and reinforces trust. Award ceremonies mean little without purpose. Real recognition defines and reinforces the example everyone should follow. Whether formal or informal, a leader's acknowledgment tells Sailors their work matters and when people feel seen, they give more. More focus, more initiative, and more loyalty to the mission and to each other. Across well-led commands, recognition lives on the deckplates. From letters of appreciation to advancement ceremonies to a well-timed Bravo Zulu at quarters, these moments create momentum. Even a quiet "well done" from a Chief or Division Officer can shift a Sailor's mindset. On ships where recognition is part of the culture, morale stays high and performance stays sharp. In contrast, leaders who fail to recognize contributions often find their teams disengaged, even if standards are met. Leadership isn't just about correcting what's wrong but about celebrating what's right.

Take the Helm
- **Give Credit Often and Authentically** – Highlight a Sailor's performance after a tough watch, a successful maintenance check, or a solid drill. It takes seconds to say "well done" but the impact can last a career.

- **Use Recognition to Reinforce Standards** – Praise publicly when Sailors set the example. Let others see what excellence looks like and know it's worth chasing.

Eyes on the Horizon
Meaningful recognition strengthens connection and morale. Build a crew that feels valued, and they'll fight harder, work smarter, and stay longer.

Day 29 – The Importance of Loyalty in Leadership

The 5 Levels of Leadership – John C. Maxwell

"People follow leaders they trust, and trust is built through actions, not titles."

Admiral Chester Nimitz exemplified the importance of loyalty in leadership throughout World War II, standing firmly behind his commanders in moments of controversy and pressure. During the Battle of Leyte Gulf, Admiral "Bull" Halsey was widely criticized for pursuing a decoy Japanese carrier force northward, temporarily leaving the San Bernardino Strait unguarded and nearly allowing Japanese surface ships to attack vulnerable American landing forces. Rather than issue a public rebuke, Nimitz addressed the situation by highlighting its complexity and reaffirming his commitment to trusting his commanders' judgment in combat. Admiral Raymond Spruance was criticized by some for his cautious tactics during the Battle of the Philippine Sea, particularly for not pursuing the retreating Japanese fleet more aggressively after the famous "Great Marianas Turkey Shoot." Nimitz, however, fully supported Spruance's decision-making, citing the need to protect the invasion force at Saipan as a valid strategic priority. His unwavering support built confidence throughout the chain of command, enabling his team to act boldly and recover quickly when things didn't go as planned. That environment of mutual trust and loyalty was a key factor in the Navy's victories across the Pacific. When leaders are loyal to their people, they earn it in return and the whole command moves forward as one.

Take the Helm
- **Demonstrate Loyalty Through Advocacy** – Speak up for your Sailors, defend their efforts, and support their development especially when no one's watching.

- **Lead with Integrity, Earn Their Trust** – Set the standard for honesty, fairness, and professionalism. When your Sailors know they can count on you, they'll go anywhere with you.

Eyes on the Horizon
Loyalty should be earned, not demanded. Build it through action and your Sailors will stand with you when it matters most.

Day 30 – The Role of Innovation in Leadership

Good to Great – Jim Collins

"The good-to-great companies did not say, 'Let's get innovative.' They said, 'What are we passionate about? What can we be the best in the world at?' And innovation followed."

Improving what matters most takes more discipline than novelty. Jim Collins reminds us that greatness comes from initiative and focused curiosity. That means questioning outdated methods, exploring better solutions, and refusing to settle for "that's how we've always done it." Innovation is strategic evolution. The best leaders foster a mindset where improvement is expected and initiative is rewarded. That way of thinking goes beyond operations and lays the groundwork for decisive warfighting dominance. At the dawn of the 20th century, most senior naval officers dismissed aviation as a passing novelty, but Admiral William Moffett saw its potential and led the charge to integrate aircraft into fleet operations. Against heavy resistance, he championed the development of the aircraft carrier and laid the foundation for carrier strike group dominance. What set his vision apart was the willingness to defy convention and the resolve to improve it. Today, U.S. naval power rests on the legacy of leaders like Moffett who saw what was possible and pursued it with purpose.

Take the Helm
- **Foster a Culture of Improvement** – Encourage your Sailors to bring ideas forward on how to improve processes, maintenance, and operational planning. Reward initiative.

- **Lead Change Thoughtfully, Not Reactively** – Use data, feedback, and operational outcomes to evolve how your team trains, fights, and sustains readiness.

Eyes on the Horizon
Innovation doesn't always mean disruption. Sometimes it simply means refusing to settle. The future belongs to leaders who evolve with purpose and lead with vision.

CHAPTER 2:
THE TRUSTED LEADER

Day 31 – Trust as the Foundation of Leadership

It's Your Ship – D. Michael Abrashoff

"The more control I gave up, the more command I got."

Trust is built through consistent behavior, credibility, and genuine care for your Sailors. D. Michael Abrashoff's experience commanding *USS Benfold* is a masterclass in this principle. He inherited a crew marked by low morale and disengagement, yet through daily presence and an open-door approach to feedback, he earned their trust and transformed performance. Rather than leaning on rank or authority, he modeled the behavior he expected and trusted his Sailors to take pride and ownership in their work. As *USS Benfold's* Sailors saw that their input mattered, they bought into the mission and into each other. The result was undeniable. *USS Benfold* achieved the highest retention rate in the Pacific Fleet and became a benchmark for operational excellence with Sailors proactively identifying problems and implementing solutions that improved readiness across the board. The secret was trust earned daily by a commanding officer who put his crew first and backed his words with action. In every division, on every ship, trust remains the foundation of cohesion and mission success. Without it, leadership crumbles into compliance and burnout, but with it, teams thrive even under the harshest conditions.

Take the Helm

- **Be Predictable in Character and Action** – When your Sailors know who you are, as someone who will show up, follow through, and support them, then they'll give you everything they have.

- **Build Trust Through Shared Ownership** – Involve your Sailors in decisions that affect them. Empowerment builds trust faster than any speech ever could.

Eyes on the Horizon

Trust doesn't mean stepping back. Empowering others begins when leaders step aside and allow growth to happen. The more you share ownership, the more your team invests in the mission and in each other.

Day 32 – The Moral Courage to Speak Truth

Daring Greatly – Brené Brown

"Clear is kind. Unclear is unkind."

Real honesty demands discomfort, especially when the easy path is staying quiet. Brené Brown reframes vulnerability not as weakness, but as the birthplace of trust, accountability, and meaningful leadership. She explains that real connection, especially in teams, depends on the willingness to be honest, clear, and courageous. Trusted leaders don't hide behind vague feedback or sugarcoat hard truths to keep the peace. They choose clarity over comfort because they care enough to challenge those around them to grow. Moral courage means telling a Sailor when they're off course, confronting toxicity in the mess, or admitting when you've fallen short yourself. That willingness to be seen and to speak plainly is a mark of strength, not fear. In the fleet, those who are trusted the most are the ones who never waver in hard conversations. During post-inspection debriefs, personnel evaluations, or even pre-deployment readiness assessments, the temptation is always there to soften the message or avoid confrontation. But integrity demands the opposite. Leaders must be willing to say the hard thing and then stick around to do the work that follows. That's what builds trust. Truth paired with a steady hand.

Take the Helm
- **Be Direct With Purpose** – Whether counseling a junior Sailor or giving feedback up the chain, speak with clarity, not avoidance

- **Lead the Hard Conversations** – Don't outsource the tough discussions. Model the courage to tackle them head-on.

Eyes on the Horizon
Truthful leadership isn't harsh. The most honest leaders speak clearly because they care deeply. When you pair honesty with support, your words become a foundation for growth.

Day 33 – Empowerment Builds Ownership

Turn the Ship Around! – L. David Marquet

"Don't move information to authority. Move authority to the information."

L. David Marquet's experience aboard *USS Santa Fe* showed that giving away control, when paired with competence and clear intent, strengthens command. When authority is pushed to the Sailors closest to the information, you get quicker decisions, sharper awareness, and stronger ownership. On any given ship, this principle plays out daily: Sailors on watch, in repair lockers, or manning consoles must be empowered to make sound decisions without waiting for permission at every turn. When leaders trust their teams to act, they build a command climate that runs on initiative, not micromanagement. Consider a scenario where a junior engineering watchstander identifies an abnormal reading and has the confidence to secure a system before it crosses a critical threshold. Without empowerment, that same watchstander might hesitate, waiting for orders while a preventable issue escalates. Over time, this undermines confidence, delays critical decisions, and weakens the very trust that mission-ready teams depend on. Instead, the watchstander acts not because they were told, but because they know they're expected to lead from their post. That's how a ship becomes more agile, more resilient, and more mission-ready.

Take the Helm
- **Train Your Sailors to Act, Not Wait** – Give them clear left and right limits and reinforce the trust to act within them.

- **Give Ownership of Processes** – Let Sailors own maintenance, training, and inspections. They'll take pride in what they manage.

Eyes on the Horizon
Micromanagement kills initiative. Empowered Sailors own the mission and they execute it better.

Day 34 – Recovering from Lost Trust

Extreme Ownership – Jocko Willink and Leif Babin

"When a leader sets the example and takes ownership, problems get solved, and the team starts to trust."

Every leader stumbles, but what defines great leadership is the willingness to own mistakes and rebuild trust through consistent action. When leaders deflect or dodge responsibility, they erode trust. And when trust breaks down, Sailors stop speaking up. Initiative fades, communication suffers, and the default becomes "just follow orders," not out of respect, but out of resignation. That silent disengagement is the most dangerous symptom of lost trust and it spreads fast. But when Sailors step forward, acknowledge failure, and commit to change, they begin the long road back to credibility. And recovery isn't a one-time apology. Real recovery comes through consistent behavior that proves the lesson stuck. In 2017, when *USS John S. McCain* collided with a commercial vessel, leading to the loss of ten Sailors, the incident exposed deep flaws in training, leadership, and operational oversight. Instead of burying the failure, the Navy accepted accountability, initiated sweeping reforms, and overhauled shiphandling and navigation training across the fleet. This deliberate and public ownership demonstrated that trust could begin to be restored, not through rhetoric, but through action and sustained change.

Take the Helm
- **Own the Mistake, Fix the Process** – If you've lost your Sailors' trust, address the issue directly, communicate the lessons learned, and lead the reset.

- **Rebuild with Consistency, Not Just Words** – Trust is restored through daily action. Be reliable, honest, and transparent in everything you do.

Eyes on the Horizon
Trust may fracture in a moment, but it's rebuilt over time by leaders who own the failure and rise with humility, strength, and renewed purpose.

Day 35 – Leading with Honor

Call Sign Chaos – Jim Mattis and Bing West

"You must always engage your brain before your mouth but never hesitate to speak the truth."

Honor in leadership means fidelity to what is right even when that choice carries personal risk. It's about living by values and making ethical decisions under pressure. Leaders who act with honor set the tone for integrity across the command. Those who cut corners, stay silent, or self-preserve in moments of adversity may never recover their credibility. When Captain Brett Crozier faced a COVID-19 outbreak aboard *USS Theodore Roosevelt*, believing his Sailors' health was at risk, he sent a message up the chain requesting urgent help. His decision cost him his command, but it earned him the respect of his crew, who recognized his willingness to lead with honor, even at personal cost. After his relief, Crozier didn't protest or attempt to preserve his reputation through the media. Instead, he bore the outcome with dignity, never wavering from the values that led him to speak out. His example shows that honor is about making the right call and standing by it, even when it costs you everything. Crozier's example reminds us that honor doesn't always align with convenience or career advancement, but it always aligns with what leadership demands.

Take the Helm
- **Live and Lead by Your Values** – When faced with gray areas or unpopular truths, choose integrity over comfort. The right decision rarely feels easy but it's always right.

- **Protect Your Sailors, Even When It's Risky** – Honor means advocating for your people, standing up when others won't, and never compromising what's right for what's easy.

Eyes on the Horizon
A uniform alone does not grant honor. Every hard choice in favor of what's right strengthens it. Do it consistently, and your Sailors will trust you completely.

Day 36 – The Power of Transparency

Leaders Eat Last – Simon Sinek

"When we feel safe, we trust. And when we trust, we will follow."

Simon Sinek recounts the story of a company during the 2008 financial crisis to illustrate the principle of transparency as a foundation of trust. Bob Chapman, CEO of Barry-Wehmiller, was faced with the need to cut $10 million in expenses. Instead of resorting to layoffs, he gathered his team and communicated the situation with full transparency, laying out the company's financial position, the risks to its survival, and the need for shared sacrifice. Chapman then proposed a solution called the "furlough program," where every employee would take four weeks of unpaid leave, but not all at once and not all together. His message was clear: "It's better that we all suffer a little than any of us suffer a lot." That act of transparency and shared burden built a deeper sense of trust and loyalty among employees, ultimately strengthening the organization through the crisis rather than fracturing it. This brutally honest environment ensured that problems surfaced early, feedback was candid, and decisions were made with full understanding of the stakes. The company emerged from the crisis financially stable and culturally stronger, proving that transparency when paired with shared sacrifice, can reinforce organizational resilience.

Take the Helm
- **Communicate With Clarity and Context** – Whether issuing a directive, explaining a schedule change, or delivering feedback, ensure your Sailors understand what's happening and why.

- **Model Honesty From the Top** – Admit when you don't have the answer, acknowledge when things go wrong, and create an environment where truth is expected, not feared.

Eyes on the Horizon
Transparency is strength, not a weakness. Lead with clarity, speak with honesty, and your Sailors will follow with confidence.

Day 37 – Leadership Through Respect

The Five Dysfunctions of a Team – Patrick Lencioni

"Respect is the foundation of teamwork."

Patrick Lencioni outlines how team dysfunction begins when members don't feel safe to speak up. That level of safety begins with a culture of respect, shown not only in recognition, but also in daily tone, fairness, and presence. Leaders who treat every team member with dignity build unity and trust. Those who interrupt, belittle, or dismiss ideas, intentionally or not, create distance and disengagement. In high-stakes environments like the Navy where trust and cohesion are critical, leadership rooted in respect becomes a force multiplier for mission success. On any Navy ship, this principle plays out in the rhythm of everyday leadership. A division officer who enforces the standard with consistency, listens to concerns without sarcasm, and treats junior Sailors with the same professional courtesy as their peers builds lasting influence. During inspections, maintenance meetings, or quarters, the tone leaders set matters. When Sailors feel seen and heard, even while being held accountable, they lean in rather than pull back. Crews like this don't just comply with orders, they commit to the mission, because they know their leadership respects them as professionals and people. That's what turns a group into a team and a team into a winning crew.

Take the Helm

- **Demonstrate Respect in Every Interaction** – Whether you're conducting field day or listening to the concerns of a junior Sailor, treat everyone with the same professionalism and regard.

- **Set the Tone Through Small Actions** – How you correct mistakes, handle disagreements, or recognize effort will shape your division's culture. Respect starts with you and spreads from there.

Eyes on the Horizon
Respect is about example. Sailors commit when they feel valued.

Day 38 – Leading With Accountability

Principles – Ray Dalio

"If you don't own your mistakes, you can't learn from them."

Ray Dalio describes how he built Bridgewater Associates into one of the most successful hedge funds in the world by fostering a culture rooted in radical transparency and accountability. At Bridgewater, every employee, from interns to executives, was expected to be brutally honest about mistakes, identify root causes, and document key learnings. One of the most striking examples is how Bridgewater created a real-time feedback tool called the "dot collector." During meetings, everyone rates each other on various attributes, like openness, assertiveness, or decision-making, in real time. These ratings were discussed openly. When someone made a poor decision or fell short, the failure was never hidden or excused. Leaders dissected it to find lessons and prevent repeats. This created a climate where accountability became celebrated as the path to excellence. Leaders who owned their mistakes were respected because their openness allowed the entire team to improve. Dalio believed that this culture of self-confrontation and collective learning was the reason Bridgewater consistently outperformed its peers in complex, high-pressure environments. In a Navy context, this translates directly: accountability strengthens cohesion. When Sailors see leaders model this kind of honesty, they follow suit, creating teams that adapt quickly, own their roles, and grow together even when they fail.

Take the Helm
- **Own Mistakes Publicly, Correct Them Quietly** – When something goes wrong, take public responsibility. Then work behind the scenes to correct, coach, and prevent.

- **Model Accountability to Instill It** – When your Sailors see you take responsibility, they'll feel empowered to do the same without fear.

Eyes on the Horizon
The most respected leaders are the ones who take responsibility without being asked. If you own it, your Sailors will follow.

Day 39 – The Role of Emotional Control in Leadership

With the Old Breed – Eugene B. Sledge

"Fear and panic are killers. Calm and discipline save lives."

In crisis, it's the leader's poise that steadies the team. Eugene B. Sledge's memoir of fighting with the 1st Marine Division in Peleliu and Okinawa provides one of the most searing examples of emotional control in leadership. Amid the mud, rain, and relentless shellfire, Sledge describes how junior leaders kept their Marines focused by mastering their own fear. In Okinawa, he recounts moments when entire units huddled in shell holes surrounded by enemy fire and death. What held them together was the steadying voice of a corporal or sergeant who refused to panic, who calmly loaded his weapon, gave short orders, and reminded others to focus on the next task. One such moment came when his unit was pinned down in a ravine and mortar rounds bracketed their position. Rather than run or scream, the squad leader scanned the terrain, picked a viable route of escape, and moved with measured urgency saving lives, not through heroics, but through composed action under chaos. In naval operations, from flight deck emergencies to combat systems casualties, that kind of presence defines a leader. Emotional control is active discipline under stress, and it becomes the model your Sailors follow when it matters most.

Take the Helm
- **Maintain Composure in High-Stakes Scenarios** – Whether you're responding to a casualty or handling conflict in your division, your ability to stay calm sets the emotional tone for the entire team.

- **Respond with Intention, Not Instinct** – Before reacting to mistakes or setbacks, pause, assess the situation, and speak with clarity. Your judgment carries more weight when it's deliberate.

Eyes on the Horizon
Your Sailors will take their cues from you. When you stay composed, they stay focused. In the heat of operations, calm leadership is what keeps the ship steady.

Day 40 – The Power of Leading by Example

Leadership in War – Andrew Roberts

"The generals who won were the ones who stayed with their men, slept in the mud, and bled when they bled."

Leadership isn't about directing from a distance. Showing up, especially in hard moments, is where leadership takes root. Andrew Roberts shows that legendary commanders like the Duke of Wellington and General Eisenhower earned lasting respect through consistent presence as much as strategic brilliance. Wellington was seen sleeping on the battlefield beside his troops, enduring the same discomfort and danger. Before the D-Day invasion, Eisenhower did not hide away in a headquarters. He visited the troops, walked among the paratroopers, asked about their lives, and shared their rations. These weren't photo ops. They were moments of shared hardship that communicated: "I won't ask you to do anything I wouldn't do myself." That truth still applies in today's Navy. Sailors constantly observe their leaders. It's the Division Officer who dons coveralls for a bilge clean, the Chief who joins a FOD walkdown at 0500, or the CO who stands on the weather decks with the team during a freezing replenishment at sea. These actions require no speech, but they speak volumes. Presence earns trust. And when it matters most, your Sailors will follow, not because you're in charge but because they've seen you lead from the front.

Take the Helm
- **Lead with Visible Discipline and Professionalism** – From standing watch to preparing briefs, your Sailors will copy your standard and raise it through action.

- **Outwork and Outprepare** – Be the first in the space, the last to quit, and the one who sets the tone for effort and accountability.

Eyes on the Horizon
Your example is your legacy. Lead with it and your Sailors will follow without hesitation.

Day 41 – Empathy Is a Leadership Strength

Daring Greatly – Brené Brown

"Empathy fuels connection. Sympathy drives disconnection."

Empathy isn't about being soft. Leading with empathy means understanding your team's reality without losing your resolve. Brené Brown emphasizes that real empathy requires courage, presence, and a willingness to understand others without judgment. Leaders who dismiss emotions or offer surface-level sympathy create distance and disengagement, but those who listen deeply, acknowledge challenges, and respond with respect foster a culture of trust. Empathy becomes a tool for readiness in the Navy's demanding environments. On any ship, empathy is tested in moments both large and small, during extended watches, last-minute schedule changes, or personal challenges happening behind the scenes. A division leader who notices a Sailor showing signs of fatigue and takes the time to ask, "How are you holding up?" is doing more than checking a box. They're building trust. Whether adjusting a workload temporarily, providing clarity in chaotic situations, or simply listening without judgment, empathetic leadership shows Sailors that they matter beyond their billet. That connection deepens commitment. Empathy creates the climate where Sailors stay mission-focused because they know their leadership has their back.

Take the Helm

- **Listen Without Judgment** – Whether a Sailor is struggling with personal issues or professional stress, hear them out before responding. Empathy is a force multiplier.

- **Lead with Compassion, Uphold the Standard** – Sailors don't need a soft leader, they need one who understands and leads them through hard times without letting go of expectations.

Eyes on the Horizon

Empathy deepens leadership by aligning strength with understanding. Lead with understanding, and your Sailors will follow with trust.

Day 42 – Leading With Integrity

The 7 Habits of Highly Effective People – Stephen R. Covey

"If you don't stand for something, you'll fall for anything."

Integrity is the anchor of trusted leadership. It means doing the right thing, especially when it's inconvenient, unpopular, or unobserved. Leaders who uphold integrity build trust, foster accountability, and set a moral tone that defines the entire command climate. But when integrity falters, it takes everything with it: credibility, morale, and cohesion. After the 2000 terrorist attack on *USS Cole*, the crew's leadership faced unimaginable pressure in the wake of tragedy. Amid chaos, grief, and uncertainty, leaders prioritized transparency, accountability, and the well-being of their Sailors. Their response wasn't improvised. Their actions reflected a command culture built on integrity long before the crisis hit. Because that trust was already established, their transparency in crisis felt authentic, not performative, and it shaped how the Navy would respond to tragedy going forward. When leaders act with integrity long before adversity strikes, they build the kind of resilient teams that respond with strength in the face of catastrophe.

Take the Helm
- **Model the Navy Core Values in Daily Action** – Lead with Honor, Courage, and Commitment whether standing watch, qualifying a Sailor, or preparing a brief.

- **Set the Standard for Accountability** – If something goes wrong, own it. Then correct it openly and professionally so your team learns what right looks like.

Eyes on the Horizon
Integrity is your most powerful leadership tool. It can't be issued, and it can't be faked. When you lead with consistency and character, trust follows and builds mission-ready teams.

Day 43 – The Danger of Favoritism

The Five Dysfunctions of a Team – Patrick Lencioni

"The moment a leader shows favoritism, resentment poisons the team."

Favoritism is one of the quickest ways to erode trust and destroy team cohesion. Patrick Lencioni explains that when team members perceive unequal treatment, whether in recognition, discipline, or access to opportunity, it triggers resentment and disengagement. Even subtle favoritism breeds dysfunction causing teams to protect themselves rather than collaborate. In high-performing environments, fairness is non-negotiable. This principle applies directly to the Navy where shipboard life magnifies every leadership decision. A junior Sailor penalized for a minor infraction that others had previously received only verbal counseling for is an inconsistency that will be immediately noticed. Leaders should quickly correct imbalance, apply fairness and affirm that expectations are applied equally, regardless of rate, seniority, or past performance. As Lencioni makes clear, when fairness becomes the norm, dysfunction fades, and unity takes its place. The standard must not shift based on who's standing in front of you because credibility, once lost, is hard to regain.

Take the Helm
- **Establish One Standard and Apply It to All Ranks** – Whether it's watchstanding, qualifications, or accountability, make sure your Sailors see that expectations apply equally across all ranks.
- **Promote and Reward Based on Merit** – Recognize Sailors who perform, lead, and uphold the mission, not just those you personally connect with.

Eyes on the Horizon
Nothing fractures a division faster than favoritism. Lead with fairness, and your Sailors will unite behind the mission, not compete for approval.

Day 44 – The Integrity to Choose What's Right

The Mission, the Men, and Me – Pete Blaber

"It's not reality unless it's shared."

Leaders are constantly choosing between what's easy and what's right. Pete Blaber emphasizes the importance of operating in a "shared reality," a condition where everyone is grounded in truth, not perception, politics, or optics. That reality begins with integrity. Having led at the highest levels, Blaber teaches that strong leadership means hunting for the real truth, accepting it fully, and taking action, no matter the discomfort. Integrity means not doctoring logs to avoid scrutiny, not glossing over inspection failures, and not letting small discrepancies snowball into operational risk. It means doing the right thing even when no one is watching or especially when everyone is. Leaders who operate this way build long-term trust even when it means taking short-term heat from their chain of command. On the deckplates, integrity shows when a junior Sailor reports a maintenance discrepancy accurately even if it risks delaying a certification or when a Department Head cancels an evolution because conditions aren't safe despite external pressure to proceed. These decisions aren't easy, but they protect crews, uphold standards, and ultimately enhance mission effectiveness.

Take the Helm
- **Refuse to Mask the Truth** – Always report the true state of readiness, regardless of external pressure or optics.

- **Make the Hard Right Call** – When standards clash with convenience, choose the harder path. It builds long-term trust.

Eyes on the Horizon
Integrity is your compass. When you act with honesty, your team learns to do the same, even under pressure.

Day 45 – Consistency Builds Trust

The Culture Code – Daniel Coyle

"Safety is not mere emotional weather. It's the foundation on which strong culture is built."

Daniel Coyle explains that trust in high-performing cultures emerges from a leader's ability to create consistent psychological safety. He studied elite organizations like SEAL Team Six, Pixar, and the San Antonio Spurs, revealing that what binds these teams together is not just talent, but a deep sense of belonging reinforced through steady, intentional leadership. One of the most important behaviors was consistent presence from leaders who show up with humility and fairness, day after day. Coyle describes how SEAL Team Six debriefs every mission with brutal honesty and zero ego, reinforcing trust with reliable patterns of feedback, ownership, and accountability. This kind of culture is built when leaders treat every interaction, whether a reprimand or a check-in, as a chance to show that standards won't fluctuate with mood or stress. A CO who walks the spaces at the same hour each night, or a Chief who listens before responding even under pressure, signals that the crew's psychological safety isn't situational, but earned and reinforced. Sailors take their cue from what leaders do consistently, not occasionally, and when leaders lead with consistency, they gain the freedom that comes from a crew who trusts them without hesitation.

Take the Helm
- **Be the Same Leader Every Day** – Whether the ship is in port or under fire, your Sailors should always know what kind of leader you'll be.

- **Apply Standards Fairly and Predictably** – Consistency in discipline, recognition, and expectations builds the foundation of a trusted command climate.

Eyes on the Horizon
True leadership is measured in moments repeated, not moments remembered. Be steady, be clear, and your Sailors will trust you with the mission.

Day 46 – The Leadership Skill of Listening

The 7 Habits of Highly Effective People – Stephen R. Covey

"Most people do not listen with the intent to understand. They listen with the intent to reply."

Listening is one of the most underrated but essential skills in leadership, especially in the Navy, where Sailors operate in fast-paced environments. As Stephen R. Covey teaches, genuine listening means setting aside ego and truly seeking to understand. Near-miss incidents on a ship's flight deck, for example, highlight how quickly operations can become dangerous when concerns are ignored or dismissed. When Sailors notice unsafe trends in procedures but aren't always confident their input would be taken seriously, then that hesitation creates risk. In contrast, commands where leaders foster open communication where even the quietest Sailor is encouraged to speak up benefit from faster issue resolution, stronger morale, and higher overall performance. Listening isn't just courteous. True listening underpins clear operations and decisive action. Problems escalate when Sailors feel unheard, but when leaders create space for honest input, without fear of dismissal or retaliation, they gain the insight they need to lead more effectively. Trust, safety, and mission success all start with listening.

Take the Helm

- **Hold Space for Feedback** – Make room in briefs and debriefs for Sailors to share insight, concerns, or corrections, then act on what's valid.

- **Listen Without Preparing a Rebuttal** – Listen with the goal of understanding, not defending, even junior Sailors have the information you need.

Eyes on the Horizon
Your Sailors will tell you what you need to know if you're a leader they believe is listening.

Day 47 – Humility Strengthens Command

Leading With the Heart – Mike Krzyzewski

"Ego and pride must be replaced by humility and purpose. That's when the team becomes more important than the individual."

Leadership is about building a team that can think and act decisively within the framework of the mission. Confidence may earn respect, but humility builds trust, and in leadership, trust is everything. Coach K led Duke Basketball to five national championships. His most powerful leadership lessons, however, came from how he faced setbacks and stayed humble amid success. When coaching the U.S. Men's Olympic Basketball Team, Coach K deliberately invited junior players to lead team meetings and encouraged input from athletes like LeBron James and Kobe Bryant, despite their fame. Rather than being performative, it demonstrated a mindset of continuous growth and genuine engagement. That humility had strategic effect. It broke down barriers between stars and staff, created buy-in, and established a culture of mutual respect. In naval leadership, this directly translates to senior leaders valuing input from junior Sailors during pre-watch briefs and training sessions. When leaders create space for others and model a willingness to grow, they forge teams that are more committed, more cohesive, and more capable.

Take the Helm
- **Admit What You Don't Know** – You don't need all the answers. You need the discipline to learn and the humility to ask.

- **Share Success, Take the Blame** – Let your team shine in victory and shield them when things go wrong.

Eyes on the Horizon
Arrogance isolates leaders. Humility connects them to their people.

Day 48 – Accountability Over Blame

Resilience – Eric Greitens

"We all fail. The question is how you respond to failure. Resilience is about finding a way forward."

Eric Greitens writes candidly about his own struggles and failures as a way to confront them, not escape them. After resigning as governor of Missouri amid scandal, Greitens did not attempt to blame political enemies or circumstance. Instead, he turned inward and wrote a series of letters to a fellow SEAL struggling with hardship using his own failings as a teaching tool. This was echoed in the U.S. Navy's response to the *USS Fitzgerald* collision in 2017. The incident claimed the lives of seven Sailors and exposed deep breakdowns in seamanship, training, and leadership. The Navy undertook sweeping reviews and reforms across the surface fleet. Leadership accepted institutional responsibility and implemented new standards in navigation training, watchstanding practices, and bridge resource management. The Surface Warfare Officer School curriculum was overhauled, new assessments, such as the Mariner Skills Logbook and increased simulator training, were put into place. In both cases, the lesson is the same: accountability isn't about blame, but about ownership that drives action. That's how trust is restored and resilience is cultivated, through leaders who own failure and turn it into forward momentum.

Take the Helm
- **Take Responsibility, Then Take Action** – If you make a poor decision in planning, execution, or communication, own it publicly, correct it quickly, and turn it into a teaching point for the team.

- **Shield Your Sailors, Sharpen Their Skills** – If your division misses the mark, own the result. Then double down on training, mentorship, and standards to prevent repeat failure.

Eyes on the Horizon
What matters most is not perfection, but the willingness to own the result. When you carry the weight of the outcome, your Sailors will trust you to carry them through the mission.

Day 49 – Consistency Builds Credibility

The Culture Code – Daniel Coyle

"Consistency of behavior is the foundation of safety and belonging."

Consistency in word and action allows Sailors to predict your decisions and emulate your leadership. Daniel Coyle explains that elite teams are built by consistent behaviors that reinforce safety and shared purpose. Take it from the U.S. Navy's Blue Angels. Their extraordinary flight precision does not come from talent alone. Cultural consistency makes it possible. After every flight, each pilot publicly debriefs with radical honesty, owning mistakes without defensiveness. This ritual never changes, regardless of rank or outcome. The message is clear: accountability is a shared standard, not a slogan, and because the behavior is consistent, so is the trust. Sailors don't need dramatic speeches. What they need are leaders whose tone doesn't change with stress, whose standards don't shift with circumstance, and whose presence remains steady when things go wrong. Consistency allows teams to focus on execution rather than guessing what kind of leader they'll get that day. When Sailors know how you'll respond, not just in ideal conditions, but in the middle of chaos, they'll trust your leadership, even when the decisions are hard. You don't have to be perfect. Just be dependable.

Take the Helm
- **Be Predictable in Your Leadership Behavior** – Don't confuse your team with shifting standards. Your steady example gives them confidence.
- **Reinforce the Basics Daily** – The best leaders revisit expectations constantly, ensuring that small wins build into long-term habits.

Eyes on the Horizon
Credibility isn't built in a moment. Show up the same every day and your Sailors will know they can count on you.

Day 50 – The Power of Ethical Decision-Making

Call Sign Chaos – Jim Mattis and Bing West

"If in doubt, always choose to do the harder right over the easier wrong."

Jim Mattis and Bing West reflect on moments when doing the right thing meant resisting pressure, accepting risk, or standing alone. Leaders who compromise values to protect image or convenience may win short-term, but they forfeit credibility and the moral authority their teams depend on. That truth was painfully illustrated after the sinking of the *USS Indianapolis* in 1945. Following a successful mission to deliver atomic bomb components, the ship was torpedoed and sank in minutes. Hundreds of Sailors perished awaiting rescue and despite acting under orders and without escort, Captain Charles McVay was court-martialed, widely seen as a political move to shift blame. Decades later, the Navy formally cleared McVay's record, and in 2000, Congress passed a resolution exonerating him, recognizing that he had been wrongfully blamed. Survivors of the *USS Indianapolis* had long advocated for his vindication and their efforts ensured that the truth prevailed. History now records McVay as a leader who was failed by the system, not one who failed his crew. His case stands as a sobering reminder that ethical leadership must be paired with institutional courage and that doing what's right may not bring immediate reward, but it will always matter in the end.

Take the Helm
- **Hold the Line on Integrity in Every Decision** – Whether you're managing evals, reporting issues, or advising up the chain, choose what's right even when it's unpopular.

- **Build a Culture Where Ethics Are the Norm** – Reward honesty. Address dishonesty. Your crew will follow your lead.

Eyes on the Horizon
Ethical leadership builds trust that endures beyond your tour. When the pressure rises, integrity is your compass. Never let it waiver.

Day 51 – Leadership Through Self-Discipline

With the Old Breed – Eugene B. Sledge

"War is brutish, inglorious, and a terrible waste. Combat leaves an indelible mark on those who endure it. But discipline, instilled and internalized, is what holds a man together when fear, fatigue, and chaos threaten to tear him apart."

Freedom, whether in action, decision-making, or effectiveness, comes rom disciplined habits and personal accountability. Leaders who lack discipline are unpredictable, inconsistent, and difficult to trust. On the other hand, those who show up early, stay prepared, and hold themselves to the highest standard build a reputation that their teams can rely on. When your Sailors see that you don't cut corners and don't ask for anything you're not willing to do yourself, they don't follow out of obligation, they follow out of respect. Aboard any ship, especially during high-tempo periods like pre-deployment workups, a leader's self-discipline can set the tone for the entire command. If you consistently arrive early, fully prepared, and locked in during drills, inspections, and morning quarters, it signals to the crew that standards matter and that leadership starts with example, not enforcement. Over time, this discipline becomes contagious. Sailors begin showing up sharper, performing cleaner, and owning their roles with more pride. When discipline is modeled from the top, cohesion strengthens and execution improves, not because of pressure, but because of trust. A disciplined leader creates the space for a high-performing crew to thrive.

Take the Helm

- **Lead Through Personal Discipline** – From fitness to watchstanding, your Sailors notice your habits. Be the example they follow.

- **Stay Sharp, Stay Ready** – When you live with discipline, you never have to scramble. Readiness becomes routine.

Eyes on the Horizon

Self-discipline earns respect before you ever speak a word. Live it and your leadership will never be questioned.

Day 52 – Clarity in Command

Leadership Strategy and Tactics – Jocko Willink

"If your team doesn't understand what you're doing or why, that's on you not them."

Confusion is a leadership failure. Jocko Willink makes it clear that when a team doesn't understand the mission, the breakdown starts at the top. Clear intent, simple communication, and shared understanding allow teams to operate effectively especially under pressure. One such situation involved a training mission where Willink gave his team vague guidance, assuming they understood his intent. As the operation unfolded, confusion mounted, team members hesitated and second-guessed one another. The lack of clear, shared understanding stalled progress and increased the risk of friendly-fire during the scenario. The team was not at fault, as Willink realized. He had failed to communicate simple, clear orders and confirm they were understood in advance. His solution from then on was to keep the plan simple, communicate it clearly, and confirm it's understood. Aboard ships, especially during complex evolutions like night flight operations or coordinated sea-and-anchor details, clarity is critical. If one department receives conflicting guidance or an ambiguous tasking, the result can be delays, friction, or worse, safety hazards. But when leaders take the time to synchronize plans, communicate intent across departments, and establish a single, unified standard, teams move with precision. A well-briefed crew operating under a clearly communicated mission executes with confidence and trust. Being clear doesn't slow things down. Clarity creates momentum by helping everyone move in the same direction.

Take the Helm
- **Always Communicate Intent, Not Just Instructions** – Let your Sailors understand the purpose of your directions so they can adjust when the unexpected happens.

- **Brief Clearly, Debrief Honestly** – Clear communication before and after the mission builds trust, learning, and better outcomes.

Eyes on the Horizon
Confusion breeds hesitation. Clarity builds speed, trust, and tactical momentum especially when the pressure hits.

Day 53 – The Balance Between Authority and Approachability

Leaders Eat Last – Simon Sinek

"People don't follow rank. They follow those they trust and respect."

True leadership is grounded in meaningful connection, not just formal authority. Simon Sinek explains that effective leaders inspire loyalty not through intimidation or formality but through trust and empathy. Leaders must protect and care for those under their charge while upholding the mission. If leaders are too distant, Sailors become hesitant to speak up or engage. The balance lies in leading with strength and humanity, commanding respect while remaining accessible. This balance plays out in how leaders walk the spaces, correct mistakes, and listen to their teams. A division officer who upholds the standard while still asking, "What do you need?" builds both trust and credibility. An approachable Chief who holds Sailors accountable *and* takes time to mentor earns respect that rank alone can't buy. When leaders are never seen outside the wardroom, Sailors feel disconnected. But when leaders are engaged, visible, consistent, and fair, the crew leans in. Authority gives you the position. Approachability gives you the influence. The best leaders know how to use both.

Take the Helm

- **Balance Authority with Approachability** – Enforce standards and make tough calls but do so with consistency and respect. Your Sailors should know where the line is and that you'll lead them, not intimidate them.

- **Foster a Climate Where Speaking Up is Safe** – From safety concerns to new ideas, create an environment where Sailors feel heard. That trust starts with your reaction the first time they come to you.

Eyes on the Horizon
Balance presence with purpose and your Sailors will give you both respect and trust.

Day 54 – The Impact of Discipline on Leadership

Make Your Bed – Admiral William H. McRaven

"Discipline is the difference between being good and being exceptional."

Admiral William H. McRaven shares that in SEAL training, countless missions succeeded not because of heroics, but because every detail had already become reflex. Without discipline, small cracks widen into mission failure. But when discipline becomes habit, it frees the mind to focus fully on the task at hand. There is no hesitation, no second-guessing, just execution. In the Navy SEAL community, discipline is non-negotiable. Their strict training, relentless attention to detail, and deep commitment to team accountability are what make flawless execution possible. McRaven recounts a mission where he and his SEAL team executed a high-risk operation with limited visibility, narrow timing, and enormous consequences. Every movement was coordinated and every command crisp. That level of performance wasn't trained overnight. Thousands of small, consistent actions made it possible. That's what discipline builds: the capacity to operate calmly in chaos. In the Navy, that might look like a junior Sailor flawlessly executing a casualty response drill during a real fire, not because they were told exactly what to do in that moment, but because they had already done it a hundred times in training. Discipline shapes leaders and sharpens the path to high performance.

Take the Helm
- **Model Discipline in Every Space You Walk Into** – Whether conducting quarters, walking spaces, or standing watch, show your Sailors what right looks like because your standard becomes theirs.

- **Never Compromise the Standard for Convenience** – Small lapses lead to bigger failures. Hold the line even when it's unpopular or inconvenient.

Eyes on the Horizon
Discipline goes beyond routine. When leaders consistently enforce the standard, they shape a culture where excellence becomes second nature.

Day 55 – Fairness is a Leadership Standard

The 5 Levels of Leadership – John C. Maxwell

"Leadership gains respect when it treats everyone with equal dignity."

John C. Maxwell recounts a moment in his leadership journey when a high-performing employee was given a pass for poor behavior while others were held to the standard. The damage wasn't immediate, but it was deep. Respect faded, resentment grew, and the team lost confidence in their leader's consistency. Maxwell emphasizes that true leadership is earned through fairness by holding everyone accountable with equal dignity, not bending expectations for convenience or preference. The same holds true in the naval environment where Sailors measure a leader not by rank or charisma, but by consistency in word and action. On any given ship, fairness shows up in the daily grind, advancement boards, watchstanding qualifications, leave chit approvals, and discipline. When one Sailor is counseled for being late to quarters and another is ignored because they're a top performer, the message spreads fast: standards are optional for some. But when leaders apply expectations equally, regardless of rate, seniority, or rapport, they send a powerful signal: this is a command built on integrity. That kind of fairness protects morale, reinforces trust, strengthens cohesion, and earns the respect that rank alone can never command. When your Sailors believe in your standard, they'll rise to meet it.

Take the Helm
- **Be the Same Leader for Every Sailor** – Whether they're the hardest charger or a work in progress, apply expectations evenly.
- **Base Recognition and Discipline on Performance** – Reward what's earned, not who you like. Correct what's wrong, not who's new.

Eyes on the Horizon
Fairness isn't optional. Be consistent, and your Sailors will know they can trust you.

Day 56 – Building Trust Through Actions

Dare to Lead – Brené Brown

"Trust is earned in the smallest of moments, not through grand gestures."

Trust is earned, not in speeches or dramatic moments, but over time through small, consistent actions. As Brené Brown reminds us, it's not grand gestures but micro-moments: how we listen, follow through, and show up that form the foundation of trust. This was evident in the *USS Johnston*, *USS Hoel*, and *USS Samuel B. Roberts* during the Battle off Samar. Commander Ernest E. Evans didn't earn his crew's loyalty the day he charged a superior Japanese fleet. He earned it beforehand, by knowing his Sailors, training alongside them, and leading with consistency. His crew trusted him because he proved daily that he'd never ask more than he gave. Across *Taffy 3*, leaders at every level showed that trust is forged long before crisis. Sailors aboard didn't follow orders blindly. They followed officers, Chiefs, and LPOs who had earned their respect through fairness, honesty, and care. When the call came to face impossible odds, their teams didn't hesitate. The decision had already been made during field day, in training, on watch that these were leaders worth following. That's the power of daily trust: small, steady actions that move Sailors from compliance to commitment.

Take the Helm
- **Follow Through Commitments, Big or Small** – Whether it's backing a Sailor for a qualification, reviewing a maintenance check, or following up on a schedule change, your credibility is built on your consistency.

- **Be Steady and Fair in Every Leadership Situation** – Hold the same standard in quarters, on watch, or during inspections. Consistency builds confidence in your leadership.

Eyes on the Horizon
Trust is the sum of your smallest decisions. Earn it daily, and your Sailors will follow you when it matters most.

Day 57 – Fairness Is Not Weakness

The Five Dysfunctions of a Team – Patrick Lencioni

"When there is an absence of trust, it stems from a failure of vulnerability."

Being fair isn't about uniform treatment. Fairness is about applying the same standard to all, with respect and transparency. Patrick Lencioni identifies inconsistent accountability as major threats to trust and team cohesion. When leaders judge personalities instead of performance, they create a culture of guardedness, under-communication, and quiet resentment, something Sailors are quick to detect in the close quarters of shipboard life. But when leaders lead with fairness, making expectations clear, decisions consistent, and consequences predictable, trust begins to form, and morale rises. Sailors watch closely: they know when recognition is earned or when it's handed out based on who's liked. They notice when one watchstander is counseled for a lapse while another, with more seniority or personal connection, gets a pass. On a ship, these inconsistencies compound quickly. Fairness shows the quiet power of disciplined integrity, keeping the division united and reminding everyone that each Sailor matters, not just those with the loudest voice or strongest ties.

Take the Helm

- **Check Your Bias at the Door** – Be aware of your assumptions and base decisions on objective performance and standards.

- **Be Transparent in Recognition and Discipline** – Let your team know the purpose behind your actions so they trust the process.

Eyes on the Horizon
Fairness fosters trust and trust multiplies performance. Your crew will follow you farther when they believe you lead without partiality.

Day 58 – The Power of Humble Leadership

Good to Great – Jim Collins

"The greatest leaders channel their ambition into the greater cause, not themselves."

Jim Collins describes "Level 5 Leaders" as those who blend professional will with personal humility, leaders who are fiercely committed to the mission but free of ego. Their humility creates space for learning, collaboration, and shared ownership. Rather than driving performance through fear or recognition, they inspire it through quiet consistency and unwavering standards. In Collins' research, it was these humble leaders, not the charismatic ones, who built the most enduring and effective organizations. In a naval environment, where operational success depends on trust, humility becomes a force multiplier. Admiral Raymond Spruance embodied this principle during World War II calmly leading from behind the scenes, favoring deliberate planning, direct communication, and mission-first execution. He sought input, gave credit, and avoided the spotlight, earning the deep respect of subordinates and peers alike. Volume did not define his command presence. Clear direction and solid skill did. Like Collins' Level 5 Leaders, Spruance's humility reflected focused strength, not passivity. In today's fleet, the most effective leaders still follow that model. Listening more than they speak, deflecting praise, and focusing every decision on what's best for the team, not their own advancement.

Take the Helm
- **Credit the Watch Team, Not Just the Outcome** – Whether it's a successful inspection, drill, or deployment milestone, highlight the Sailors who made it happen and let higher leadership know their names.

- **Seek Input from All Ranks** – From junior enlisted to seasoned Chiefs, listen actively. The best decisions often come from the deckplates, not the wardroom.

Eyes on the Horizon
Humble leaders don't seek the spotlight. They stay anchored to purpose and keep the mission front and center. Put your Sailors first, and they'll make you a leader worth following.

Day 59 – Trust Is Built in Failure

Grit – Angela Duckworth

"Enthusiasm is common. Endurance is rare."

One moment of adversity doesn't define grit. Grit is forged through endurance, sustained effort in the face of repeated setbacks, and refusal to quit. Angela Duckworth defines grit as passion and perseverance over the long term especially when success doesn't come quickly. It's not about avoiding failure but showing up again and again after things go wrong. For leaders, this consistency builds trust. Sailors trust leaders who don't just recover from one mistake but who continue to lead with transparency and determination after every misstep or disappointment. Whether it's a division that fails an inspection, a repeated discrepancy that resurfaces, or a string of miscommunications during a deployment, these are opportunities to demonstrate grit. A leader who takes ownership, corrects the course, and stays focused on the mission, even after multiple setbacks, teaches the team that failure is part of growth, not the end of it. Over time, that attitude becomes contagious. Sailors begin to internalize the idea that perseverance is the standard, not the exception. And that's when trust takes root, not in flawless execution, but in the unwavering resolve to keep getting better.

Take the Helm
- **Own Mistakes and Share Lessons** – Model the humility and transparency that teaches your team how to respond after setbacks.
- **Debrief With Courage** – Create a culture where failure is analyzed honestly, not hidden or feared.

Eyes on the Horizon
The foundation of trust isn't perfection, but the courage to be accountable when it matters most. When your Sailors see you lead through failure, they'll follow you with greater conviction.

Day 60 – The Legacy of a Trusted Leader

The 5 Levels of Leadership – John C. Maxwell

"Your legacy will be defined by the leaders you leave behind."

Even after the orders are signed and the nameplate is changed, the influence of a trusted leader lives on in the people they've prepared to lead. John C. Maxwell teaches that the highest form of leadership isn't found in authority, but in the legacy you leave behind. The greatest leaders invest in others, shape command culture, and build systems that continue to thrive in their absence. They lead with a long view, developing the character and capability needed to endure well beyond the present mission. This principle is seen clearly in leaders like Admiral Hyman Rickover, whose rigorous standards created a culture of nuclear excellence that still defines the submarine force today. Admiral Chester Nimitz empowered commanders like Spruance and Halsey, fostering trust and unity across the Pacific Fleet during World War II and leaving behind a generation of operationally minded naval leaders. And MCPON Delbert D. Black, the Navy's first Master Chief Petty Officer of the Navy, helped institutionalize enlisted leadership and set the foundation for the senior enlisted structure that still guides the fleet. By investing in people and principles, these leaders created enduring legacies that still shape naval culture today.

Take the Helm
- **Lead with the Future in Mind** – Every policy, decision, and relationship contributes to the legacy you leave. Make yours one that strengthens the command.

- **Grow Leaders Who Will Surpass You** – Train, coach, and develop your Sailors to carry on the mission without you. Your impact is measured by how they lead next.

Eyes on the Horizon
You won't wear the uniform forever but the standard you set can live on in every Sailor you lead. Legacy is built through example every single day.

CHAPTER 3:

LEADERSHIP AS A PROFESSION

Day 61 – Leadership Is a Profession, Not a Position

The 5 Levels of Leadership – John C. Maxwell

"Leadership is not about titles, positions, or flowcharts. It is about one life influencing another."

John C. Maxwell outlines how leaders grow through influence, not rank. He explains that the lowest level of leadership is position and the highest is legacy where others lead well because of the example you set. The greatest leaders take the craft seriously. They reflect, learn, and strive for consistent self-improvement. In the Navy, leadership must be treated like any profession with study, humility, mentorship, and repetition. When Sailors embrace this mindset, they move beyond simply managing tasks. They begin shaping teams, building culture, and creating lasting impact. That mindset is embodied in the legacy of the Chiefs' Mess. A device on a uniform does not grant true influence. Guiding Sailors, modeling technical skill, tackling hard talks, and mentoring are what leave a lasting mark. Chiefs model what it means to treat leadership as a profession. One that is deliberate, accountable, and mission focused. Their daily investment in growth and guidance proves what Maxwell teaches: leadership at its best is a life's work and not just a line on a chart.

Take the Helm
- **Treat Leadership Like a Skill, Not a Badge** – Study, reflect, and grow every day. The best leaders never finish learning.

- **Lead Through Daily Influence** – Make your presence felt not by rank, but by how you invest in others, consistently and deliberately.

Eyes on the Horizon
Leadership is a profession. Study it, live it, and pass it on. When you treat it as a craft, your influence outlives your position.

Day 62 – Discipline Begins with Perspective

The Obstacle Is the Way – Ryan Holiday

"Where the head goes, the body follows. Perception precedes action. Right action follows the right perspective."

Ryan Holiday reminds us that our mindset is the starting point of disciplined action. He draws from Stoic philosophy to show that the way we interpret adversity shapes how we respond to it. Leaders who see setbacks as fuel for growth and pressure as an opportunity to lead develop the resilience and mental clarity needed to perform under stress. This mental discipline, the ability to remain calm, focused, and forward-thinking amid chaos is the foundation of professional leadership. Holiday recounts the story of Ulysses S. Grant during the Civil War, who once entered a house under artillery fire and sat calmly to write dispatches while shells exploded nearby. His officers and staff were visibly shaken, some even diving for cover, but Grant remained unflinching. His reaction was a conscious decision to master fear through perspective. His calm under fire reassured his men, steadied morale, and reinforced his authority as a leader who would not yield to panic. As Holiday explains, Grant's example showed that when leaders control their response, they create stability for others. His mindset multiplied discipline across the entire team.

Take the Helm

- **Train Perspective Before Pressure Hits** – Help your Sailors see challenges as opportunities to grow, adapt, and lead.

- **Model Calm in the Chaos** – Your mindset sets the tone. Mental discipline is contagious and critical in crisis.

Eyes on the Horizon

Professionalism under pressure starts with the discipline of how you think. Your team will take their emotional cues from you. Lead with steady perspective, and you'll create clarity in chaos.

Day 63 – Lead Yourself First

Dare to Lead – Brené Brown

"Who we are is how we lead."

Before you can lead others, you must first develop the courage and self-awareness to lead yourself. Brené Brown emphasizes that grounded leadership begins with character. This means doing the uncomfortable work of self-reflection, owning your behavior, and aligning your actions with your values. This translates to leading with quiet consistency. The Sailor who shows up early, upholds standards, and follows through without fanfare earns lasting trust. No one embodied this more than Master Chief Carl Brashear, the Navy's first African American Master Diver. After losing his leg in a devastating accident, Brashear didn't seek sympathy. He committed himself to full recovery and requalification, setting a standard that left no room for excuses. His resilience and refusal to let adversity define him became a masterclass in self-leadership. Whether it's a division officer preparing meticulously before briefs or a Chief setting the example in daily inspections, these leaders lead from within. Their strength comes from the discipline to live the standard first. Self-leadership isn't a performance but a promise that inspires others to follow.

Take the Helm
- **Be the Baseline** – Be the Sailor who keeps qualifications current, double-checks maintenance logs, and follows up on tasking without reminders. Your habits set the baseline for your division.

- **Your Readiness Sets the Pace** – Hold yourself accountable to the highest standards during qualifications, inspections, and drills. Your example during these moments will shape how your team views responsibility and readiness.

Eyes on the Horizon
Before you can ask others to follow you, show them who you are through your actions. In a profession built on trust, self-leadership is where credibility and command begins.

Day 64 – Commit to Lifelong Learning

Mindset – Carol S. Dweck

"Why waste time proving over and over how great you are, when you could be getting better?"

In a naval environment where missions evolve and technologies advance, growth-minded leaders stay sharp through study, curiosity, and humility. Nowhere is this more evident than in the story of Boatswain's Mate 1st Class James E. Williams, who rose from humble beginnings to become one of the most decorated enlisted men in Navy history. He didn't do it through ego or flash, but through tireless effort, tactical learning, and operational mastery. While patrolling the treacherous rivers of Vietnam, Williams constantly adapted to new threats, refined his judgment, and applied lessons learned in real time. His success was the product of a leader who treated every patrol like an opportunity to grow sharper and more capable for the team. That same mindset is what turns a good Sailor into a trusted leader. Lifelong learners seek out doctrine, policies, and mentorship not for promotion, but for performance. They improve their planning, refine their judgment, and elevate those around them because they see leadership not as a destination, but as a craft to be mastered.

Take the Helm

- **Invest in Your Professional Edge** – Dedicate time each week to professional reading, whether it's leadership development, warfare doctrine, or Navy policy updates, and apply what you learn to improve your team, your watchstanding, and your planning.

- **Reflect, Refine, Repeat** – After each major evolution, drill, or inspection, conduct a personal debrief using your notes or logs. Assess what went well, where you fell short, and how to apply those lessons moving forward.

Eyes on the Horizon

Leadership is a journey. Stay curious, stay humble, and keep learning. The best leaders in the fleet are never finished products. They remain students of the craft, always improving so they can better serve their Sailors and the mission.

Day 65 – Pass Down What You've Learned

Leaders Eat Last – Simon Sinek

"Leadership is not about being in charge. It is about taking care of those in your charge."

Simon Sinek emphasizes that the best leaders build trust by serving those they lead, and one of the greatest forms of service is mentorship. Experience, when kept to yourself, creates distance, but when passed down, it builds connection, confidence, and continuity. It's essential, not extra, to mentor others through qualifications, failures, and personal growth. This culture of mentorship is modeled every day by Command Master Chiefs across the fleet. Often viewed as walking archives of experience, they guide Sailors not just through policy, but through the hard-earned lessons of leadership, adversity, and accountability. Whether guiding a new LPO through a difficult conversation or standing by a Sailor after NJP, these leaders choose to lead with investment. Judgment has no place in their approach. Their credibility comes not from position, but from a career of showing up, passing it on, and making others better. Their legacy lives not only in what they accomplished but in the people they developed. The strength of the team depends on how well we prepare the next watch to take the helm.

Take the Helm
- **Share What You've Learned** – Incorporate real-life fleet examples into your training sessions. Share stories of past successes and failures to connect the lesson to real operational consequences.

- **Make Mentorship a Habit** – Use downtime during watches, work center meetings, or after-action reviews to pass down rate-specific knowledge, tactical insights, or leadership lessons that accelerate your Sailors' development.

Eyes on the Horizon
The knowledge you share today builds the leaders of tomorrow. Your willingness to mentor and teach doesn't just strengthen your division, it shapes the next generation of Sailors.

Day 66 – Hold the Line on Standards

Call Sign Chaos – Jim Mattis and Bing West

"A leader must always be ready to take the initiative and enforce standards."

Standards are not suggestions. Treating standards as non-negotiable protects trust, guarantees safety, and keeps the force ready. Leaders who let things slide create a culture where mediocrity takes root. Upholding the standard requires moral courage: the willingness to correct lapses, even when it's uncomfortable, and to reinforce what right looks like in every evolution, watch, and interaction. That principle was exemplified aboard *USS Porter* after its 2012 collision in the Strait of Hormuz, a mishap that revealed a breakdown in accountability and command climate. In the wake of the incident, leadership changed, training intensified, and standards re-established from the deckplates up. What followed was a remarkable turnaround: within a year, *USS Porter* regained combat readiness, re-earned the trust of the fleet, and received accolades for operational performance. The ship's recovery was driven by leaders who refused to compromise, reinforced discipline daily, and proved that culture can be rebuilt one standard at a time. Their actions echoed Jim Mattis' and Bing West's guidance: the mission starts with leadership, and leadership starts with standards.

Take the Helm

- **Uphold the Standard Without Exception** – Set clear, achievable expectations during quarters, training, and evolutions then enforce them with consistency and fairness to maintain trust and operational integrity.

- **Correct Early, Consistently, and Fairly** – Use formal counseling, documentation, and mentorship sessions not just to correct performance, but to guide Sailors back on course while reinforcing the standard and preserving their ownership in the process.

Eyes on the Horizon

Professionalism is built on standards. Protect them and your team will thrive. When you enforce expectations with fairness and follow-through, you create a culture where Sailors know what right looks like and are proud to uphold it.

Day 67 – Set the Example Even When It's Hard

The Dichotomy of Leadership – Jocko Willink and Leif Babin

"If you want to be a good leader, you must do the things you expect your team to do. You must lead them by example."

Chief Gunner's Mate Paul Foster exemplified this principle during a gunnery exercise aboard *USS Trenton* in 1923. When a shell ignited and threatened to detonate the forward magazine, Foster, already a seasoned leader responsible for his team's safety didn't seek cover or wait for direction. Instead, fully aware of the mortal danger, he acted decisively and flooded the magazine, preventing a catastrophic explosion. His action, which cost him his life, was the choice of a leader who knew that setting the example meant owning the hardest decisions in the worst moments. Foster didn't ask anyone to do something he wasn't willing to do himself. That standard, leading through sacrifice, left a legacy of courage that continues to inspire Sailors a century later. Today, when an LPO demonstrates a task during a stressful moment instead of barking orders or a department head stays behind to help clean up after an evolution, it tells the crew that standards apply at every level. When Sailors see you lead the hard way, they'll follow you the same way.

Take the Helm

- **Show Up When It's Tough** – Don't wait for ideal conditions. Be present, prepared, and engaged when the stakes are high or the task is unpleasant. Your readiness is a signal that standards apply to everyone, including the leader.

- **Be the Standard, Not Just the Voice** – When mistakes occur, take ownership immediately. This builds credibility and teaches accountability by example.

Eyes on the Horizon

The best leaders lead from the front. Your actions in quiet, unrewarded moments will define the culture of your team more than any words ever could.

Day 68 – Know the *Why* Behind the *What*

Start With Why – Simon Sinek

"Working hard for something we don't care about is called stress. Working hard for something we love is called passion."

Simon Sinek explains that a successful company like Apple inspires loyalty because it leads with a clear sense of purpose: to challenge the status quo and empower individuals through innovation. This *why* is embedded in everything Apple does, from marketing to product design to customer experience, so their employees advance a cause and not just build computers. Sinek highlights that Apple employees are emotionally invested because they believe they're creating tools that help people think differently, work creatively, and live more freely. As a result, they willingly work long hours, tackle complex problems, and stay committed because they want to contribute to something meaningful. In the Navy, there's always a mission to complete or a task to check off but professional leaders go deeper by explaining not just what needs to be done, but why it matters. The more clearly you connect the task to the mission, the more your team brings care, not just compliance. These moments reframe the work. Once your team understands how their actions influence safety, readiness, and mission success, they stop just doing the job and start owning it.

Take the Helm
- **Explain the *Why*** – During briefs, evolutions, and training, clearly explain how each task supports the ship's mission, readiness, or Sailor development. Purpose fuels motivation and precision.

- **Inspire Ownership** – Show your team how their work matters. Pride follows clarity.

Eyes on the Horizon
Professionalism begins with purpose. When Sailors understand the mission behind the task, they commit, bringing focus and pride to every detail.

Day 69 – Take Ownership of Your Development

Can't Hurt Me – David Goggins

"The only way we can change is to be real with ourselves."

No one is more responsible for your growth than you. David Goggins issues a blunt challenge to anyone coasting through life: if you're not uncomfortable, you're not growing. Lieutenant Commander Jonny Kim didn't start his career as a doctor, an officer, or an astronaut. He started as a Sailor. An enlisted hospital corpsman who volunteered for BUD/S, Kim earned his place as a Navy SEAL. But that was just the beginning. With a relentless desire to grow, Kim went on to earn a degree in mathematics, graduate from Harvard Medical School, and ultimately become a NASA astronaut. He didn't wait for the Navy to decide what he could become. He owned his gaps, sought the next challenge, and pushed himself to exceed every limit that stood in his way. That's what real development looks like, not just ticking boxes for promotion, but reshaping what's possible through discipline and personal ownership. Whether you aim to lead a team, shape policy, or explore space, the principle is the same: no one will invest more in your potential than you. The question is: how far do you want to go?

Take the Helm
- **Lead Yourself Before You Lead Others** – Professional credibility begins with how seriously you treat your own advancement.

- **Build Your Own Growth Plan** – Track your accomplishments, lessons learned, and personal goals in a professional journal or leadership binder. Approach your development with the same discipline you would a mission plan or watch turnover.

Eyes on the Horizon
No one owes you success. Take ownership and earn it every day. The most respected leaders do not wait passively for growth. Intentional effort and relentless follow-through drive their progress.

Day 70 – Treat Every Role Like It Matters

Good to Great – Jim Collins

"Real leaders are modest, willful, and fearless."

Jim Collins highlights that the best organizations are built on individuals who take pride in their work regardless of their title. Warfighting readiness doesn't rest on one rate or one rank, but built by every Sailor who logs gear correctly, fixes a leaking valve, updates a tracker, or prepares meals for the watch team. Commands that excel do not leave Sailors guessing about their importance. Leaders demonstrate respect for each role, Sailors meet that standard with care in every task, and the result is a united, proud, and high-performing crew. Across the fleet, some roles get more recognition than others, but the best performing ships are the ones where every division feels essential. For example, the scullery team that cleans trays with consistency enables the galley to serve hot chow on time, a Logistics Specialist who tracks parts accurately ensures engineering doesn't stall during a casualty, and an admin clerk who routes evaluations with care protects Sailors' careers. These roles may not grab headlines, but they power the ship. When Sailors own their roles, no matter how small, they build the mindset that shapes leaders, strengthens teams, and wins wars.

Take the Helm

- **Lead From Where You Are** – Every task is a chance to lead with pride and professionalism. Own your space.

- **Respect Every Role** – Recognize and highlight outstanding performance in often-overlooked roles during quarters, spot awards, or evaluations. Reinforce that every billet contributes to mission success.

Eyes on the Horizon

There are no small jobs, only small mindsets. When you show that every contribution matters, you build a culture where Sailors take ownership in everything they do, from the flight deck to the scullery.

Day 71 – Be the Leader Others Want to Work With

The Mission, the Men, and Me – Pete Blaber

"Trust is not given. It is earned. Every. Single. Day."

Only by leading with dignity and professionalism, and by prioritizing Sailor development, can you foster a team built on mutual respect. Fleet Master Chief Paul Kingsbury understood this better than most. He made it his mission to prepare junior Sailors for leadership. Recognizing a gap in how trust and influence were taught to junior Sailors, he authored *The Petty Officer's Guide* as a call to action. His message was clear: if you want to lead effectively, you need to be the kind of leader others actually want to follow. That means showing up, walking the decks, being dependable, mentoring consistently, showing interest in your Sailors' growth, making time to listen before leading, and never forgetting that trust is your most valuable currency. Pete Blaber's message aligns perfectly: trust can't be demanded. It must be earned every single day. Rank might give you authority, but leadership that lasts comes from professionalism. Kingsbury didn't inspire high performance by issuing orders. His impact came from investing in people, shaping culture, and proving that the best leaders elevate everyone around them.

Take the Helm
- **Lead With Presence and Respect** – Routinely walk your spaces, engage your Sailors during watch rotations or maintenance, and ask how things are going. Build trust through presence, not just position.

- **Be the Leader You'd Follow** – Be consistent, fair, and transparent when making decisions, especially during high-stress situations or disciplinary moments. Your team will remember how you led when it counted most.

Eyes on the Horizon
Trust is your strongest leadership tool. Earn it with humility and consistency. The more your Sailors believe in your character and fairness, the more willing they'll be to follow you into challenge, change, or combat.

Day 72 – Stay Humble as You Rise

Ego is the Enemy – Ryan Holiday

"Impressing people is utterly different from being truly impressive."

You lose sight of the profession the moment you believe leadership is about you. Sailors at every level face the temptation to lead from ego, especially after a promotion or new assignment. Promotion changes things. Not only is it a new title, but it's also a new test. With added authority comes a subtle temptation to believe you've arrived. To stop asking questions. To stop listening. The Navy's record tells the story clearly. Each year, multiple commanding officers are relieved for "loss of confidence" and it's rarely about technical failure. More often, it's about ego creeping in, leaders growing distant, inconsistent, or dismissive of feedback after promotion. They forget what got them there: humility, discipline, and connection. What built trust erodes when ego begins to outweigh presence. That's why the most respected leaders stay grounded, they stay teachable, they continue to walk the deckplates, ask questions, and invite accountability. Leadership is a responsibility, not a reward, that demands even more humility the higher you go. It's knowing that you're still learning and always will be. Rank may mark progress, but it doesn't guarantee wisdom. The moment you stop seeking growth is the moment you start slipping from it.

Take the Helm
- **Let Performance Speak** – Earn respect through action, not image. Your team will follow your example.
- **Stay Teachable at Every Rank** – Ask for feedback often, especially after promotions, and surround yourself with people who tell you the truth.

Eyes on the Horizon
The best leaders don't let ego speak for them. When you lead with humility, you create a culture where learning, growth, and shared success matter more than rank or recognition.

Day 73 – Develop the Next Watchstander, Not Just the Next Sailor

The Leader's Bookshelf – Admiral James Stavridis and R. Manning Ancell

"Reading is the gymnasium of the mind, but mentoring is the forge of leadership."

Reading builds your perspective, but mentorship builds your legacy. Admiral James Stavridis and R. Manning Ancell highlight that while intellectual growth is vital, leadership comes to life when we invest in developing others. That means your real influence is measured not by what you achieve personally, but by how well your Sailors lead, decide, and perform when you're not in the room. True mentorship isn't just about helping Sailors get qualified but about teaching them how to think, act, and lead when it counts. Stavridis practiced this throughout his career, from ship command to flag-level roles. He took time to mentor junior officers and enlisted leaders, whether through book discussions, after-action reviews, or simply being available for tough conversations. He shared experience, teaching not just procedures, but professional judgment. As a combatant commander, he was known for developing leaders who could operate independently and ethically in fast-moving situations. That commitment to shaping leaders became his most enduring legacy. Stavridis made it clear that preparing the next watchstander is core to leadership, not an added burden.

Take the Helm
- **Build Independent Leaders** – Use watchbill assignments strategically. Rotate motivated Sailors into more challenging roles, and provide clear expectations, coaching, and feedback to help them succeed.

- **Mentor with Purpose** – Deliberately match junior Sailors with trusted mentors in your division to guide their technical proficiency, professional conduct, and leadership mindset.

Eyes on the Horizon
Your legacy isn't just what you lead but who you prepare to lead next. By developing others, you multiply your impact and build a team that thrives long after you've moved on.

Day 74 – Let the Standard, Not Your Mood, Lead the Team

The 33 Strategies of War – Robert Greene

"To lead others out of confusion, you must first be disciplined in your own mind."

Sailors don't need emotional swings. In uncertain moments, they need steady hands to maintain course. During an engineering casualty or a detect-to-engage event in combat, your response sets the tone. If you stay composed and issue clear orders, your team performs with confidence, but if you react emotionally and second-guess the plan, you introduce doubt and delay. And every second counts. Admiral David Farragut understood this during the Battle of Mobile Bay in 1864, when Union forces hesitated as their lead ship struck a torpedo and the entire formation risked collapse under enemy fire. Farragut climbed the rigging and famously ordered, "Damn the torpedoes, full speed ahead!" It wasn't recklessness. He led with confidence and a commitment to the standard of action over fear. His calm decisiveness kept the fleet moving forward and secured a critical victory. Like Farragut, the best leaders don't let stress dictate their presence. They hold the line, focus on what needs to be done, and trust the standard to carry the team forward.

Take the Helm

- **Let Discipline Guide the Moment** – Base your decisions and expectations on published standards, checklists, and command guidance. This gives your team clarity and consistency, especially during high-tempo or high-stress operations.

- **Project Stability, Not Surprise** – When something goes wrong, keep your posture, tone, and timing predictable. Your team should never have to guess how you'll respond.

Eyes on the Horizon

Professionalism means your team can trust your leadership even on your worst day. When you lead with steadiness and structure, your Sailors gain the confidence to perform under pressure because they know their leader won't crack under it.

Day 75 – Sharpen the Blade

The 7 Habits of Highly Effective People – Stephen R. Covey

"Sharpen the saw means preserving and enhancing the greatest asset you have—you."

Leadership is not fire and forget. Becoming a leader is a lifelong pursuit, a craft you continually refine. Stephen R. Covey's seventh habit, "Sharpen the Saw," reminds us that sustainable leadership depends on consistent renewal: mental, physical, spiritual, and social. For Sailors, this means carving out time for reflection, learning, and fitness even during a high-tempo deployment. Admiral James Stavridis understood this better than most as he kept a disciplined habit of reading, often beginning his day by studying leadership, history, and strategy. He believed that a leader's mind must be sharper than any tool in the arsenal and that belief led him to author *The Leader's Bookshelf*, a project inspired by his conversations with senior military leaders about the books that shaped their thinking. This series continued with *The Sailor's Bookshelf* and *The Admiral's Bookshelf* all underlining the same principle that reading is not a pastime, but a professional obligation for those entrusted to lead. By curating and sharing that knowledge, Stavridis turned his personal growth habit into a force-multiplier for others. His example reminds us that sharpening the blade is how great leaders stay sharp enough to shape others.

Take the Helm
- **Set Time for Renewal** – Block time each week for professional reading or journaling. Treat it as mission critical, then share insights with your team to multiply the value of your own growth.

- **Turn Downtime Into Development** – Use transit, duty nights, or quiet moments to feed your mind. Small investments in learning add up over time.

Eyes on the Horizon
A dull blade won't cut it when the pressure rises. Keep yours sharp through disciplined, intentional self-investment.

Day 76 – Repetition Builds Mastery

Multipliers – Liz Wiseman

"Multipliers are leaders who look beyond their own capability and focus on expanding the capacity of those around them."

The most effective leaders aren't just masters of their own tasks. The strongest leaders develop mastery in others through repetition, challenge, and trust. Liz Wiseman describes "Multipliers" as those who stretch their teams, not by micromanaging, but by creating space for learning through action. This doesn't mean throwing people into the fire without support. Growth happens through repeated reps that build confidence, judgment, and capability. This looks like senior Sailors who step back just enough to let junior Sailors lead drills or let qualified watchstanders run a casualty response. Admiral Arleigh Burke, long before becoming Chief of Naval Operations, was known for training his crews relentlessly, allowing junior officers to run war games and critique their own performance. That trust created a command culture where competence grew quickly and mastery was expected. Burke turned personal excellence into collective strength by developing those he led. In our Navy, repetition builds mastery, and the strongest leaders use it to build capability in others. The leaders who leave the deepest mark are the ones who grow capability, not just complete tasks.

Take the Helm
- **Teach Through Reps** – Let junior Sailors take the lead in drills or briefs, then walk through lessons learned to reinforce growth.

- **Build Thinking Teams** – Don't just give orders. Ask questions that require judgment and pattern recognition to strengthen decision-making.

Eyes on the Horizon
Mastery doesn't come from lectures. True mastery comes from doing. The more you develop others through reps, the more your team becomes a force multiplier.

Day 77 – Clarity Comes from Principles

Principles – Ray Dalio

"If you're not clear about the principles that guide your decisions, you risk making them emotionally and inconsistently."

Great leaders don't rely on guesswork when pressure mounts. What steadies a leader in crisis is the groundwork of clearly defined values. Ray Dalio instilled a culture of radical transparency and principles-based decision-making at Bridgewater Associates, one of the world's largest hedge funds. He required that key meetings and decisions be recorded and reviewed, so that lessons could be extracted and applied to future situations. This practice ensured that decisions were made based on consistent principles, not on mood, politics, or guesswork. Dalio believed that writing down your principles created clarity not just for yourself, but for your entire team, allowing everyone to operate from the same foundation. Over time, this approach created an environment where trust, accountability, and high performance were possible even under intense pressure. In high-stakes environments like the Navy, that clarity becomes a stabilizing force. Codified principles, whether shared in a standing order, watch turnover, or divisional philosophy, become a source of stability when everything else speeds up.

Take the Helm
- **Write Down Your Leadership Code** – Define three to five personal principles that guide how you lead your division. Demonstrate them through your actions.

- **Use Principles to Train Others** – When mentoring or correcting, tie your feedback to core values. It reinforces consistency and clarity.

Eyes on the Horizon
Purposeful leadership begins with knowing what you stand for before the pressure hits. The more clearly you define your leadership standards, the more others can rely on you.

Day 78 – Find the Lesson in Every Failure

Grit – Angela Duckworth

"Failure is not a permanent condition."

Angela Duckworth reminds us that grit isn't about avoiding failure. Grit is about how you respond to it. In the Navy, setbacks are inevitable: inspections will fall short, equipment will break, and leaders will miss the mark. What defines professionalism is not perfection but the willingness to learn, adjust, and keep moving forward. Failure becomes valuable the moment you choose to extract meaning from it and when leaders respond with honesty and humility. They create a culture where Sailors aren't afraid to fail and become committed to growing. After the 2013 grounding of *USS Guardian*, the Navy uncovered that outdated hydrographic charts contributed to the incident. Instead of dismissing it, leadership used the event to drive concrete reforms: improvements in digital chart accuracy, revised navigation planning procedures, and enhanced pre-mission risk assessments for transits in environmentally sensitive waters. It became a fleet-wide lesson in preparation, chart validation, and environmental awareness. Tragedy became catalysts for systemic learning because leaders at every level refused to waste the pain.

Take the Helm

- **Model Grit** – Own your mistakes and keep showing up with purpose. Your resilience becomes the standard.

- **Normalize Reflection** – Foster a professional environment where your Sailors can openly discuss mistakes during post-evolution reviews. Focus on lessons learned and how to apply them moving forward, not on assigning blame.

Eyes on the Horizon
Every failure has a lesson. Find it, own it, and lead forward. Your willingness to grow through adversity will set the tone for a resilient team that views setbacks not as shame but as stepping stones to excellence.

Day 79 – Stay Ready So You Don't Have to Get Ready

Great by Choice – Jim Collins and Morten T. Hansen

"The only mistakes you can learn from are the ones you survive."

Great organizations prepare with discipline long before the crisis arrives. That means you don't wait for an inspection, deployment, or casualty to flip the switch. Professional leaders stay ready. Nowhere was this more evident than aboard *USS Enterprise* during the Cuban Missile Crisis in 1962. As tensions with the Soviet Union escalated, *USS Enterprise* received orders to deploy as part of a rapidly forming naval blockade. Unlike other units that scrambled to get ready, *USS Enterprise* was already there. Her aviation squadrons were trained, her systems fully maintained, and her crew rehearsed for both conventional and nuclear contingencies. Within hours, the carrier was under way with a full air wing, integrated with supporting ships, and postured for high-tempo operations. The ship's readiness didn't happen overnight nor was it a result of last-minute cramming. It was the product of a command climate where every drill mattered and every qualification was treated like it would be needed tomorrow. And when the call came, they didn't have to play catch up. They simply stepped up to the plate.

Take the Helm
- **Set the Standard Daily** – Schedule regular pre-checks and internal assessments before inspections, certifications, or major evolutions. Treat readiness as a continuous state, not a last-minute reaction.

- **Rehearse for Reality** – Train your team to meet real-world mission demands, not just inspection criteria. Emphasize warfighting capability, procedural integrity, and confident execution.

Eyes on the Horizon
Excellence begins long before anyone is watching. Stay ready, and you'll never fall behind. When readiness becomes your standard, not your response, you lead a team that performs confidently in both evaluations and combat.

Day 80 – Practice What You Preach

Start With Why – Simon Sinek

"There are leaders and there are those who lead."

Simon Sinek teaches that leaders do more than issue directions. They guide by modeling the right actions with honesty and consistency. He emphasizes that people don't follow titles. They follow those whose actions align with their values. Sailors quickly detect inconsistency. If a leader enforces standards they don't meet, or preaches accountability while dodging responsibility, credibility fades. But when words and actions match, when leaders live the expectations they set, they earn trust, respect, and real influence. That alignment is what separates positional authority from true leadership. Vice Admiral Nora Tyson set the tone commanding Carrier Strike Group 2 through professionalism and poise, rather than showmanship. She consistently modeled the competence, calm, and team-first mindset she expected from her command, earning deep respect without needing to demand it. She not only became the first woman to command a carrier strike group but also the first woman to lead a Navy ship fleet as commander of Third Fleet. Practicing what you preach, as Tyson reminds us, is leadership Sailors remember and willingly follow.

Take the Helm

- **Be First to Step In** – Show up early, take responsibility, and meet the standard before you ask others to.

- **Be the First to Meet the Mark** – Never issue guidance or enforce standards you're not willing to follow yourself. Be the first to show up, gear up, and execute to the expectations you set.

Eyes on the Horizon
Your actions are your leadership so make sure they match your words. Credibility comes not from what you say, but from what your team sees you do.

Day 81 – Be a Student of the Profession

Learning War – Trent Hone

"The Navy's success stemmed from an institutional commitment to continuous learning."

Trent Hone shows that the Navy's success in World War II wasn't luck. Victory came from years of deliberate investment in professional education and tactical experimentation. From the interwar years through the Pacific campaigns, the Navy fostered a culture where learning was constant and adaptation was expected. Not only did that mindset shape doctrine, but the path to victory was also influenced by it as well. Fleet problem exercises in the 1920s and 1930s gave rise to real-time adjustments in carrier employment, amphibious tactics, and command and control. Leaders like Admiral Nimitz, Admiral Spruance, and others absorbed lessons from every iteration refining procedures, testing ideas, and building intellectual muscle long before the first shots were fired. When war came, they weren't caught off guard. Their ability to out-think and out-maneuver the enemy was forged through years of reading, gaming, and operational analysis. Hone makes it clear: the leaders who treated naval warfare as a craft worth mastering were the ones best prepared to win it.

Take the Helm
- **Learn Before the Fight** – Select one warfare area, leadership topic, or technical system each quarter to study in depth. Use pubs, doctrine, and fleet case studies to sharpen your operational understanding beyond the basics.

- **Study Your Craft** – Incorporate leadership lessons and recent case studies into divisional training, wardroom discussions, or CPO mess forums. Reinforce that continual learning is a professional expectation, not a personal option.

Eyes on the Horizon
The most capable leaders are the most committed learners. Study the profession and sharpen your advantage. Knowledge sustains readiness, and when you lead by learning, you raise the standard for everyone around you.

Day 82 – Professionalism Requires Self-Control

Man's Search for Meaning – Viktor E. Frankl

"The one thing you can't take away from me is the way I choose to respond to what you do to me."

Frankl reminds us that we still have a choice even in moments of pressure or chaos. Your ability to stay calm, think clearly, and act with discipline sets the tone for your team. Lieutenant Thomas Hudner lived this principle during the Korean War. After witnessing his wingman, Ensign Jesse Brown, crash-land behind enemy lines, Hudner made a split-second decision to crash his own aircraft to try and save him. Though it appeared impulsive, Hudner's choice was grounded in a clear, values-driven calculation: if there was any chance to save his wingman, he would take it and he executed that decision with calm precision, not emotional panic. As the freezing winds of Chosin Reservoir howled and enemy forces closed in, Hudner worked methodically to free Brown from the wreckage never giving in to despair. His Medal of Honor was earned with bravery but also through control and compassion when others might have frozen. Whether it's a Boatswain's Mate correcting a line-handling error during a mooring evolution or a Chief de-escalating an argument in the shop, the leaders who stay composed maintain both safety and cohesion. That presence builds credibility and it trains Sailors to do the same when their turn comes.

Take the Helm

- **Pause Before Reacting** – In moments of frustration or uncertainty, use deliberate pause-and-breathe techniques before responding. Your emotional control will anchor your team during stressful situations.

- **Lead With Poise** – Lead critiques, debriefs, and corrective conversations with firmness and professionalism. Deliver accountability without raising your voice or losing composure.

Eyes on the Horizon

Self-control is strength. When your Sailors see stability in your response even under pressure, they learn to model the same control in their own leadership and performance.

Day 83 – Don't Just Check the Box, Own the Standard

Helmet for My Pillow – Robert Leckie

"Fear ran down my back like a cold drop of rain. I was afraid because I had forgotten something—something important … And I knew that if I died, it would be because I had forgotten."

Before stepping off on a combat patrol on Guadalcanal, Robert Leckie was hit with a sudden wave of dread, not from the enemy, but from a sense that he had forgotten something essential. He scanned his gear: cartridge belt, grenades, canteen, rifle. Everything was there, yet the feeling persisted. It was the realization that even the smallest oversight could cost him his life or endanger his unit. That moment revealed the brutal truth of war: survival often hinged on discipline, preparation, and unwavering attention to the standards that others might overlook. You don't check a valve alignment, equipment tag-out, or casualty response drill just to meet a requirement. You do it because lives depend on it. When Sailors internalize this mindset, they don't treat checklists as tasks. Checklists transform into instruments of precision and warfighting readiness. Whether standing a watch in CIC or maintaining a critical system, the goal is to ensure it's done right, every time. Because the margin for error may be thinner than you think.

Take the Helm
- **Own the Process** – During PMS, evolutions, and watchstanding, don't settle for box-checking. Verify each step, ask why it matters, and use those moments to mentor your team on how their work supports mission readiness.

- **Speak Up When the Standard Slips** – If you notice a step skipped, a shortcut taken, or a process misunderstood, stop and correct it. Protecting the standard protects your team and the mission.

Eyes on the Horizon
Professionalism means taking ownership of the work and delivering it with care, every single time. By reinforcing attention to detail and pride in performance, you create a team that understands every task matters.

Day 84 – Take Feedback Like a Pro

Leadership and Self-Deception – The Arbinger Institute

"When we see others as people, we hold ourselves accountable to how our actions affect them and we become open to correction."

The Arbinger Institute teaches that the greatest barrier to growth is self-deception. When leaders justify their own behavior while discounting the input of others, they close themselves off from the very feedback that would make them better. Professional leaders invite critique, not to defend themselves, but to understand how their actions impact the team. The Navy has formalized this principle through Command Climate Assessments using the Defense Equal Opportunity Climate Survey. Commands are expected to review the results, brief the findings transparently, and develop actionable improvement plans. When done right, they show Sailors that their voices matter and that leadership is listening. But the impact of these surveys depends on what leaders do next. If a DEOCS identifies gaps in trust, fairness, or communication, it's not a mark of failure, but an opportunity to lead with credibility by addressing those concerns head-on. When leaders model transparency and a willingness to improve, they create a culture where feedback is expected and not feared.

Take the Helm
- **Make Yourself Easy to Coach**– After major evolutions, watch turnovers, or inspections, actively request feedback from your chain of command, peers, and junior Sailors. Listen without defensiveness and capture notes to identify areas for growth.

- **Model Feedback in Action** – Demonstrate how to receive and apply feedback during team debriefs or training reviews. Normalize the process so your Sailors view feedback as a leadership tool, not a threat.

Eyes on the Horizon
Feedback offers perspective that sharpens performance. And when leaders treat it as a tool for growth, the whole command levels up. Professionalism means being teachable especially when it's uncomfortable.

Day 85 – Teach What You Know

The War of Art – Steven Pressfield

"The more important a call or action is to our soul's evolution, the more resistance we will feel."

Steven Pressfield teaches that one of the best ways to overcome resistance is by sharing what you've learned. Whether guiding Sailors through their first qualification board or mentoring through a complex evolution, leaders who invest time in teaching help ensure the success of their team. Admiral Elmo Zumwalt lived this mindset. As the youngest Chief of Naval Operations in history, he challenged tradition and prioritized mentorship by publishing his now-famous "Z-Grams," direct messages to the fleet designed to teach, inform, and empower. Many resisted his reforms, but Zumwalt saw past the pushback and used his platform to build understanding and growth across every level of the Navy. Many of the reforms Zumwalt introduced, such as eliminating outdated grooming standards and advocating for equal opportunity, laid the foundation for policies that still influence Navy culture today. Pressfield's message is clear: resistance to teaching or sharing knowledge often stems from self-doubt, but pushing through that resistance by teaching creates growth for both the leader and the team. The more you share, the more you solidify your understanding and improve the collective strength of your command. Leaders who teach what they know help foster a culture of mentorship where learning and improvement are part of the daily routine.

Take the Helm

- **Commit to Mentorship** – Lead informal training during maintenance or watch rotations. Use real-world experience to turn routine tasks into learning opportunities for junior Sailors.

- **Share Knowledge** – When signing off PQS or conducting OJT, explain the reasoning behind procedures, not just the steps, so Sailors develop true understanding and confidence in their responsibilities.

Eyes on the Horizon

When you invest in passing down your experience, you multiply your impact and strengthen warfighting readiness across the fleet.

Day 86 – Write It Down, Own the Process

The Toyota Way – Jeffrey K. Liker

"Without standards, there can be no improvement."

In his book, Jeffrey Liker notes how in the Toyota production system, every action, no matter how small, is standardized, written down, and tracked. That's how they achieve consistency, identify gaps, and improve over time. The same principle applies in the Navy. Good leaders write down training goals, maintenance plans, inspections, and Sailor progress. This is not bureaucracy but a commitment to clear and deliberate leadership. Admiral Hyman Rickover knew this better than anyone. As the architect of the nuclear Navy, he demanded exhaustive documentation, written procedures, and personal accountability at every level. He believed you couldn't trust a system that only lived in someone's memory. You had to see it, verify it, and refine it. Liker explains that Toyota's success depends on standardized work documentation and problem-solving that's recorded and reviewed. Professional leaders who track performance, annotate lessons learned, and organize their workflow lead with discipline. They don't guess. They plan, adjust, and improve with precision. Writing it down doesn't slow you down. Putting it on paper helps sharpen your focus.

Take the Helm

- **Document Everything** – Develop and maintain written tools such as binders, trackers, and digital logs to manage qualifications, maintenance schedules, training records, and inspection readiness in your division.

- **Lead with Discipline** – Encourage your Sailors to keep personal notebooks for jotting down lessons learned, qualification notes, and leadership insights. Reinforce that documentation is part of owning their development.

Eyes on the Horizon

Professionalism leaves a paper trail. Write it down and lead with precision. Clear records, thoughtful notes, and organized systems reflect a leader who plans ahead, tracks performance, and never leaves success to chance.

Day 87 – Never Stop Qualifying

Atomic Habits – James Clear

"You do not rise to the level of your goals. You fall to the level of your systems."

James Clear explains that success is built on habits. Qualifications form the foundation of readiness and credibility, and professional Sailors go beyond the minimum by continuing to qualify. Each new qual expands your versatility, sharpens your skills, and increases your value to the team. Collecting signatures is not the point. Developing a steady habit of growth is. That mindset is at the heart of the Enlisted Leader Development continuum to ensure Sailors have a clear, repeatable path of growth at every level of their careers. Clear's concept of identity-based habits fits seamlessly into the Navy's qualification mindset. It's understood that leadership excellence doesn't just appear at the khaki level. From the first day aboard, habits of qualification, commitment, and continuous learning shape the leadership excellence that carries into higher ranks and beyond. A Sailor committed to qualification as a way of life moves forward on their own, always ready for the next milestone. Whether it's learning the responsibilities of another rate, qualifying in the next-level watch position, or picking up another warfare pin, these Sailors lead by example. Their systems are built for growth and they bring their teams with them.

Take the Helm
- **Build Qualification into Your Routine** – Actively pursue a qualification outside your primary watchstation or rate. Demonstrate initiative and increase your versatility as a contributing member of the command.

- **Lead with Growth** – Design divisional training plans that emphasize steady qualification progress. Reinforce that growth and readiness go beyond meeting minimum manning requirements.

Eyes on the Horizon
Every qual you earn increases your value. Never stop sharpening your edge. The more you know, the more you contribute and the more you lead by example in a Navy that rewards readiness and initiative.

Day 88 – Keep Your Standards Without Being Told

The Mission, the Men, and Me – Pete Blaber

"The common denominator of success is character."

Professionalism is measured and defined in the quiet, unobserved moments: the tag-out you double-check, the log entry you verify, the maintenance step you complete even when it would be easier to skip it. Those small, unseen choices shape your credibility. Boatswain's Mate 1st Class James E. Williams demonstrated this standard during a river patrol on Oct. 31, 1966, deep in hostile territory along the Mekong Delta. When his two-boat patrol unexpectedly encountered a vastly larger force of Viet Cong guerrillas operating from fortified riverbanks, Williams could have waited for backup or withdrawn under the rules of engagement. Instead, without orders and fully aware of the risk, he launched an aggressive counterattack, maneuvering through narrow canals for over three hours while coordinating gunfire and directing air support. His decision was about refusing to leave his Sailors exposed. No one told him to stay and fight. But his instinct, honed by discipline and deep care for his crew, drove him to take action that ultimately neutralized over 50 enemy combatants and saved countless lives. He held the standard because others were counting on him. That's character in action

Take the Helm

- **Lead When No One's Watching** – Complete checklists, tag-outs, and log entries with precision even when no one is observing. Your consistency in the small details reinforces trust in the bigger ones.

- **Do It Right Every Time** – Recognize and highlight moments when Sailors do the right thing without prompting. Use these examples during quarters or training to reinforce that integrity is always mission relevant.

Eyes on the Horizon

Character is what you do in the quiet moments. When integrity becomes your standard, Sailors will follow your example even when the pressure is high and no one is watching.

Day 89 – Lead Like It's Your Craft

With the Old Breed – Eugene B. Sledge

"War is brutish, inglorious, and a terrible waste. Combat leaves an indelible mark on those who are forced to endure it."

Eugene B. Sledge reminds us that real leadership is found in how leaders carry themselves day after day. It's about professionalism under pressure. It's about knowing your job, preparing your people, and holding the standard no matter the conditions. Master Chief Carl Brashear lived this. After losing his leg in a diving accident, he refused to accept medical retirement. Instead, through years of painful rehab, relentless training, and absolute discipline, he became the first amputee to return to full duty as a Navy diver. Brashear trained for over a year to walk again with a prosthetic leg, rising at 0400 daily to condition himself beyond medical expectations. He insisted on passing every Navy diving qualification without waiver, demonstrating that leadership means mastering your craft even when the odds, and the system, are against you. His return wasn't about proving a point. The goal was to honor the craft of leadership and show his team what excellence, resilience, and pride truly look like. In every rating and rank, leaders who treat their duties as a craft, maintaining logs, training their reliefs, checking the details no one else sees, earn the lasting respect of their teams. Sailors trust leaders who show up prepared, composed, and deliberate because when things fall apart, they need someone whose professionalism never does.

Take the Helm
- **Uphold Quiet Excellence** – Let your preparation, technical knowledge, and follow-through do the talking.

- **Make Professionalism Contagious** – Show that seriousness and pride in your duties can define the culture of your division.

Eyes on the Horizon
Craftsmanship in leadership is built through daily practice. And how you do the small things defines how you'll handle everything else.

Day 90 – Leadership Is a Career-Long Commitment

The 12 Rules for Life – Jordan B. Peterson

"Compare yourself to who you were yesterday, not to who someone else is today."

Leadership isn't a moment. How you think, how you act, and how you show up is where leadership takes root. Jordan B. Peterson reminds us that growth is personal, not competitive. It's not about outpacing someone else or chasing the next milestone, but about improving who you are, step by step, over the course of your career. Admiral Jeremy "Mike" Boorda embodied that same trajectory. Joining the Navy at age 16, he became the first Sailor in Navy history to rise from the enlisted ranks to become Chief of Naval Operations. Boorda never tried to match the resumes of his peers. He built his credibility through humility, competence, and genuine care for Sailors. Every role he took on was an opportunity to improve, to learn, and to better serve the fleet. Beyond the rank, this story is about a lifelong devotion to leading well through wins, setbacks, and everything in between. Constantly remind yourself that leadership is a life-long commitment to growth and selfless service, defined by the example you set.

Take the Helm

- **Keep Growing** – End each week with a short self-assessment. Review your performance, identify areas for growth, and set one actionable goal for the next week.

- **Lead Beyond the Title** – Keep mentoring and developing others at every stage. Show your team that leadership is about growing daily and helping others do the same.

Eyes on the Horizon
Leadership isn't a phase, but rather a personal pursuit of excellence. The toughest competition you'll face is the one in the mirror, day after day. Stay committed, stay humble, and lead in a way that outpaces who you were yesterday.

CHAPTER 4:
YOUR MORAL COMPASS

Day 91 – Do What's Right Even When It's Hard

Call Sign Chaos – Jim Mattis & Bing West

"If you must lose, lose your honor last."

Ethical leadership means choosing what's right even when it comes at personal cost. Jim Mattis and Bing West remind us that honor must outlast all else, especially in moments of pressure or failure. Few stories reflect the weight of honor more than that of Admiral Mike Boorda. In 1996, a media inquiry questioned the legitimacy of two "V" devices he had previously worn on service medals, devices meant to denote valor in combat. Though Boorda had already removed the devices nearly a year earlier, believing in good faith that he had once been authorized to wear them, he feared the controversy would damage the Navy's reputation. Rather than allow even the appearance of impropriety to reflect on the institution he loved, Boorda took his own life, leaving behind notes expressing his concern that he had brought dishonor to the service. His death stunned the fleet and underscored the invisible weight senior leaders carry. Boorda's legacy, though deeply tragic, stands as a sobering lesson: to lead with honor is to accept full accountability and to carry it with humility even when the cost is unbearable.

Take the Helm

- **Own the Outcome** – When you make a mistake, whether during a brief, inspection, or evolution, take ownership immediately and outline the correction. Your accountability sets the tone for your team.

- **Stand for Integrity** – Use divisional training or quarters to talk about integrity. Share real-world examples that show honor is non-negotiable in leadership.

Eyes on the Horizon
Ethical leadership means standing tall, especially when it would be easier to look away. When you lead with integrity in hard moments, your Sailors learn to do the same.

Day 92 – Honor Begins With Telling the Truth

Crucial Conversations – Kerry Patterson, Joseph Grenny, Ron McMillan, Al Switzler

"Speak the truth and do it with respect."

When the stakes are high and opinions vary, silence or aggression often take over, but true leaders speak the truth clearly and respectfully even when the pressure is on. The investigation that followed the catastrophic fire aboard *USS Bonhomme Richard* in 2020 revealed painful failures in readiness, maintenance, and accountability. Failures that didn't emerge from a single event, but from a pattern of unspoken truths and missed opportunities. It was the leaders who chose to speak candidly, despite the weight of scrutiny, who helped the Navy begin to heal and improve. Crucial conversations happen at every level, from junior Sailors challenging broken processes to senior leaders reporting hard truths up the chain. The *USS Bonhomme Richard* fire showed how silence and avoidance can allow problems to compound, while honest, respectful dialogue can drive real reform. The leaders who participated openly in the investigation, rather than minimizing the damage or protecting reputations, demonstrated what ethical leadership looks like and in doing so, they lived the principle that trust and honor are sustained not by image, but by courage in communication. Speaking up is rarely easy but it is always essential.

Take the Helm
- **Speak with Honesty** – When providing updates to the chain of command, whether during CASREPs, watch turnovers, or inspections, report issues accurately and without omission.

- **Break the Silence Early** – Don't wait for a crisis to speak up. Open dialogue prevents failure and drives lasting improvement.

Eyes on the Horizon
Truth builds teams, safeguards mission readiness, and preserves your credibility. The habit of honesty in small things prepares you for integrity in moments that truly matter.

Day 93 – Hold the Moral Line

The 33 Strategies of War – Robert Greene

"A leader must always lead from a place of moral clarity."

Robert Greene reminds us that in times of conflict, internal or external, a leader's greatest weapon is moral clarity. The battlefield of leadership is often subtle, filled with ethical gray zones where compromise is tempting, and the path of least resistance is appealing. That battlefield can take the form of overlooking misconduct, staying silent about inappropriate behavior, or justifying unethical shortcuts in the name of mission success. Greene emphasizes that when the fog of pressure surrounds a command, it's the leader's unwavering ethical stance that provides direction. During the Fat Leonard scandal, which unfolded publicly in 2013, while many individuals fell prey to bribery and personal gain, others held the moral line. They refused gifts, questioned suspicious activity, and reported what they knew, even when it meant risking their career. That courage reflects Greene's idea of the "counterbalance" strategy, holding firm with disciplined values that resist erosion under pressure. Those officers who refused to bend didn't just protect themselves, they upheld the integrity of the Navy and helped restore public trust in the institution. In every decision, leaders either reinforce the standard or weaken it.

Take the Helm

- **Lead with Moral Clarity** – If you observe unethical behavior, fraudulent documentation, hazing, or misuse of resources, report it through the proper channels including your chain of command.

- **Resist the Slide** – Don't justify small compromises. One step off course invites a pattern that can erode your entire command climate.

Eyes on the Horizon
When you protect the standard, you protect the service. Personal integrity lays the groundwork for successful missions and trust that spans the fleet.

Day 94 – Lead With Moral Courage

The Unforgiving Minute – Craig Mullaney

"The easy wrong is seductive. The hard right is demanding."

Moral courage means doing what's right even when it puts you at personal risk. Craig Mullaney reminds us that real leadership does not rest in convenience or popularity. Hard choices, quiet resistance, and an unwillingness to compromise under pressure reveal true character. In the Navy, those choices often come without warning. During World War II, Chief Petty Officer Charles Jackson French rescued wounded shipmates after their ship was attacked, tying a line around his waist and swimming for hours through shark-infested waters to tow a life raft to safety. No one told him to do it. There was no expectation of recognition. He simply acted because it was right. Courage, as leaders like French demonstrate, comes from the choice to act boldly when many would stand still. Whether you're confronting injustice, raising a safety concern, or putting yourself in harm's way to protect others, moral courage is the clearest signal of integrity in action. Leadership is tested in the moments where the hard right is the only right.

Take the Helm
- **Choose the Hard Right** – When confronted with difficult situations, don't settle for comfort or convenience. Let principle guide your actions.
- **Act with Courage** – Lead from the front by intervening when jokes, conduct, or behavior cross the line. Silence is complicity. A quiet correction or a well-chosen word can shift an entire culture. Don't underestimate your impact.

Eyes on the Horizon
Moral courage may not always be rewarded but it's always remembered. The strength to stand alone today sets the example for the leaders who come after you.

Day 95 – Build Trust by Keeping Your Word

The 7 Habits of Highly Effective People – Stephen R. Covey

"Trust is the glue of life. It's the most essential ingredient in effective communication."

Your credibility grows every time you do what you say you're going to do. Even something as routine as following through on a promise to fix a repeated discrepancy reinforces that your word is your currency. Leaders who keep their promises, especially the small ones, build reputations that endure. Admiral Arleigh Burke led with that kind of integrity throughout his career. His Sailors trusted him not because he gave stirring speeches, but because when he said he'd act, he did. One clear example was his advocacy for the development of guided missile destroyers and the Polaris missile program. Despite resistance from traditionalists, Burke promised to push modernization forward and he delivered. His support directly led to the Navy's strategic shift into missile warfare and the commissioning of the Farragut-class destroyers, a visible sign that when Burke committed to advancing the fleet, he followed through with action, not just words. That same standard applies today, whether it's a Leading Petty Officer who follows through training a junior Sailor, an officer who delivers on commitments before a due date, or a Chief who ensures a Sailor's concern doesn't get lost in the chain of command. In every case, follow-through creates a climate of trust that strengthens the entire team.

Take the Helm
- **Follow Through Always** – If you commit to a Sailor, whether it's career counseling, qualification support, or taking an issue up the chain, follow through quickly and reliably.

- **Let Your Word Carry Weight** – During briefs or inspections, set clear expectations and deliver on them. Your consistency sets the tone for your division.

Eyes on the Horizon
Trust is built one commitment at a time. Small promises kept lead to big credibility when it matters most.

Day 96 – Make Integrity Non-Negotiable

Fearless – Eric Blehm

"True character is revealed when no one else is around to see it."

As a cryptologic technician and special operations linguist, Senior Chief Shannon Kent's contributions often took place behind the scenes. No awards ceremonies, no headlines, just quiet excellence in the shadows. She volunteered for high-risk deployments because she believed in the mission, the people beside her, and the standard she held herself to. After her death in Syria in 2019 during a special operations mission, the Navy and broader military community publicly honored her sacrifice and service. She was posthumously awarded the Bronze Star with Combat "V," and tributes poured in recognizing her as one of the most respected special operations enablers in the field. It was only then that many learned of the depth of her impact, a reminder that integrity doesn't go unnoticed forever, even if recognition comes after the final act of service. That same standard is reflected in the small, unseen decisions we make like checking systems off the clock, logging accurately when no one's reviewing your work, or fixing a discrepancy because it's right and not because someone is watching. That same kind of moral instinct guided Kent's career. When Sailors commit to integrity in the quiet moments, they form the foundation of trust that holds a team together when the pressure is on.

Take the Helm

- **Lead Yourself First** – Your actions in solitude define your credibility in public. Do what's right even when no one will ever know.

- **Uphold Quiet Standards** – Reward Sailors who take quiet, corrective action on their own. It builds a culture of trust and accountability.

Eyes on the Horizon

The measure of your integrity isn't how you perform under inspection. How you behave when nobody's watching says more about your integrity than any formal check ever could. Silent professionalism is the strongest form of leadership.

Day 97 – The Harder Path Builds Stronger Leaders

The Obstacle Is the Way – Ryan Holiday

"Easy choices, hard life. Hard choices, easy life."

Leadership is often defined by tough decisions no one wants to make. Those decisions might involve halting a rushed evolution, reporting your own mistake, or challenging a system others quietly accept. Admiral William Sims understood this when he bypassed the chain of command to report outdated gunnery practices directly to President Theodore Roosevelt. At a time when Navy gunnery training was dangerously outdated, Sims saw firsthand how American ships lagged behind their foreign counterparts. Junior in rank and frustrated by institutional inertia, he took an extraordinary step bypassing the chain of command and writing directly to Roosevelt to expose the Navy's inefficiencies. It was a career-defining gamble. Sims could have been relieved. Roosevelt chose instead to make him the Navy's Inspector of Target Practice. His act of bold integrity led to sweeping reforms in naval gunnery and warfighting readiness, directly strengthening the fleet before World War I. Sims rejected the comfort of silence and conformity. His choice was grounded in truth, professional standards, and a deep sense of service. Like every Sailor faced with a tough call, the right decision may not win immediate applause. In time, though, it builds the kind of leader others trust.

Take the Helm

- **Prioritize Principles Over Pressure** – When pressured to overlook a procedural step, safety check, or reportable issue, take the harder road, stop, verify, and escalate if needed.

- **Disrupt with Purpose** – Teach your Sailors that courage doesn't always mean charging ahead. Sometimes, it means having the strength to delay or correct course.

Eyes on the Horizon
Leadership isn't about smooth sailing. Leadership means facing rough waters and guiding through hard decisions with unwavering clarity and courage. The hard right today protects your team and your credibility tomorrow.

Day 98 – Admit Mistakes Without Excuses

The Unforgiving Minute – Craig Mullaney

"The hardest part of leadership is being honest with yourself when no one else will be."

Owning your mistakes is leadership in its rawest form, not a weakness. This means standing tall when something goes wrong and focusing on solutions. When leaders show they're willing to take the hit, their teams learn that honesty matters more than image. Craig Mullaney recounts a devastating episode in Afghanistan that tested his moral and professional judgment as a young officer. Acting on intelligence that enemy forces were using a village near the Pakistan border as a staging area, he authorized a nighttime mortar strike. The next morning, he discovered that civilians had been killed, an outcome that shattered any illusion of clean decision-making in war. Mullaney didn't rationalize the mistake or shift blame. He reported it, owned it, and lived with the consequences. He describes the haunting silence that followed, the sleepless nights, and the internal reckoning that reshaped how he led going forward. Rather than becoming defensive or numb, he emerged more reflective, more deliberate, and more committed to the human weight of command knowing that real leadership means standing accountable when no one else can or will.

Take the Helm
- **Lead with Accountability** – When your decision or oversight leads to a mistake, own it immediately, brief it honestly, and lead the recovery.

- **Build a Culture of Ownership** – Create a division culture where mistakes are addressed constructively, not hidden. Set the tone that honesty is the expectation.

Eyes on the Horizon
Owning your mistakes won't weaken your reputation. Strength of character becomes more visible with every honest admission. And integrity grows when leaders choose accountability over ego.

Day 99 – Loyalty to Mission, Not Just to People

Leadership Strategy and Tactics – Jocko Willink

"If you allow your loyalty to an individual to cloud your judgment, you could fail the team."

Jocko Willink warns that blind loyalty to individuals, traditions, or peer groups can compromise effectiveness and erode trust. True loyalty sometimes means making unpopular decisions like holding a peer accountable or challenging a long-standing norm. That kind of loyalty isn't comfortable, but it's what sustains integrity and long-term mission success. Admiral William Pratt demonstrated this as CNO in the early 1930s. Amid economic depression and global uncertainty, he supported the London Naval Treaty, which imposed limits on naval armament. Many within the Navy opposed the treaty, fearing it weakened maritime strength, but Pratt believed in the broader strategic objective, stabilizing international power dynamics and aligning with U.S. diplomatic policy. Rather than aligning with popular opinion, he stayed loyal to operational readiness, forward-thinking innovation, and the belief that peace is secured through strength and partnership. He held firm under peer criticism because he knew the Navy didn't just need more ships. What the fleet needed was focus, discipline, and forward-looking leadership. That decision, while contested, proved him loyal to the institution's future, not just its traditions.

Take the Helm
- **Hold the Line Ethically** – If a peer, superior, or tradition undermines operational effectiveness or morale, have the courage to speak up. Integrity demands that your loyalty serves the mission first.

- **Lead with Mission-First Loyalty** – Base your decisions on what's right for the command, not on who might be affected. Fairness is a form of leadership.

Eyes on the Horizon
True loyalty honors the mission, the standard, and the Sailors who rely on you to lead with integrity. Courage means choosing what's right, not just what's easy or familiar.

Day 100 – Integrity Is Leadership Under Pressure

Leadership in War – Andrew Roberts

"The moral quality of leadership is never more vital than when things are going badly."

Leadership is truly tested and revealed in moments of chaos, failure, and doubt, not when everything is going right. Andrew Roberts profiles leaders like Winston Churchill, Ulysses S. Grant, and Admiral Horatio Nelson, commanders whose integrity under pressure defined their legacy. Churchill endured criticism and isolation during the darkest days of World War II yet refused to bend his values for political comfort. Grant, facing immense casualties in the Civil War, accepted responsibility for failure and continued to fight with steady resolve. These leaders didn't hide from pressure, they stood inside it with moral clarity. Roberts shows that the difference between ordinary and great leadership is found in how leaders respond when everything unravels. Admiral Nelson, even in his final battle at Trafalgar, remained calm, honest, and focused on the welfare of his men. He led not with bluster, but with conviction proving that strength and humility are not opposites. In the Navy today, the same principle applies. What defines integrity is not flawlessness but the willingness to be honest, own mistakes, and stand firm when it matters most. When operational setbacks occur, when fatigue sets in, or when the spotlight turns harsh, your example must hold. That's when your team will learn what leadership truly means.

Take the Helm

- **Stay Grounded in Crisis** – In stressful moments, slow down and stay grounded in truth. Honest assessments are the starting point for real solutions.

- **Tell the Truth When It's Hard** – Teach your team that owning mistakes and reporting truthfully is expected, even when consequences follow.

Eyes on the Horizon

Integrity means telling the truth even when it costs you. In chaos, character steadies the ship and defines the leader.

Day 101 – Be Honest About What You Don't Know

Mindset – Carol S. Dweck

"No matter what your ability is, effort is what ignites that ability and turns it into accomplishment."

Leadership doesn't require having all the answers. What matters more is the humility to admit what you don't know and the discipline to seek the truth. Carol S. Dweck teaches that growth happens when we embrace challenges, stay curious, and stop trying to prove we're always right. Rear Admiral Grace Hopper lived this philosophy. A pioneer in computer science and naval innovation, she was known not just for her brilliance, but for her relentless willingness to learn and question. She warned, "The most dangerous phrase in the language is: 'We've always done it this way.'" Hopper challenged norms, asked questions others avoided, and believed that curiosity was a form of courage. Her success wasn't built on pretending to know everything, but on showing others that learning is a leader's most powerful tool. Like Dweck, Hopper believed that becoming something better was always more powerful than staying still. When leaders admit what they don't know, they open the door for collaboration, problem-solving, and credibility. Honesty becomes a leader's most reliable tool for earning respect and driving results.

Take the Helm
- **Own the Unknown** – When asked a question you can't answer, admit it, research it, and follow up. Your honesty will build trust
- **Lead with a Questioning Attitude** – Encourage your team to ask questions and seek clarity. Foster an environment where learning is valued over pride.

Eyes on the Horizon
You don't have to know everything. You just have to be honest and committed to learning. Humility in knowledge earns trust in leadership.

Day 102 – Enforce Standards Fairly

Leaders Eat Last – Simon Sinek

"Integrity is when our words and deeds are consistent with our values."

Integrity is more than personal honesty. When Sailors see rules bent for some but not others, discipline fades and resentment grows. But leaders who enforce expectations equally, regardless of rank, relationship, or past performance, establish a culture where fairness and professionalism can thrive. Admiral William Fallon demonstrated that commitment to integrity at every level of command, culminating in his historic appointment as the first naval officer to lead U.S. Central Command. In a role traditionally held by Army or Air Force generals, Fallon brought a balanced, mission-focused approach rooted in fairness and ethical consistency. His calm, even-handed leadership style was defined by clear standards and impartial accountability, ensuring that decisions were driven by merit and performance, not politics or personal loyalty. By leading with principle in one of the military's most complex and politically sensitive commands, Fallon proved that integrity is not just a personal virtue, but a strategic asset that builds unity, credibility, and enduring success.

Take the Helm

- **Apply the Standard to All** – Apply discipline and corrective action based on behavior, not billet or popularity. Your fairness will define your leadership.

- **Lead with Unshakable Integrity** – Model accountability by admitting when you fall short of standards. Set the tone that no one is exempt.

Eyes on the Horizon

Fairness is a force multiplier. When everyone knows the rules apply equally, your team gains confidence in you and in each other.

Day 103 – Say No to the Shortcut

Helmet for My Pillow – Robert Leckie

"There are no shortcuts to the things worth doing right."

Shortcuts might feel efficient, but they erode discipline. In war, shortcuts don't save time. They cost lives. Robert Leckie writes candidly about the brutal lessons learned as a Marine in the Pacific campaign where discipline wasn't glamorous. It was muddy, exhausting, and repetitive. But every time someone skipped a step, failed to clean a weapon, ignored a field order, or took an easier path, it put the entire unit at risk. Survival demanded habits, not hacks. That same mindset applies in every Navy workspace, from the flight deck to the engineering plant. In 1989, a devastating explosion aboard *USS Iowa* killed 47 Sailors when a turret detonated during a gunnery exercise. The tragedy exposed procedural gaps and lapses in safety protocols, reminders that discipline must govern every evolution, no matter how routine. When shortcuts are tolerated, the consequences can be deadly. Cutting corners in maintenance, training, or safety might feel efficient in the moment, but it undermines trust and invites failure. There's nothing heroic about shortcuts. What matters is the consistent work that holds the line under pressure. In the end, it's not the easy route that prepares you to lead, but the one you took the right way, every time.

Take the Helm
- **Respect the Process** – Train your team to recognize when skipping steps undermines safety, readiness or integrity, and empower them to stop the process.

- **Choose the Hard Right** – When pressured to "make it work," stand firm. Do it by the book and explain why to your Sailors.

Eyes on the Horizon
Shortcuts cut more than time. Every shortcut comes at a cost and that cost is often trust. Hold the line on standards and your team will learn to do the same.

Day 104 – Stand Up for Those Who Can't

Daring Greatly – Brené Brown

"Integrity is choosing courage over comfort."

Leadership is about conviction. Brené Brown reminds us that integrity means standing up, especially when it's uncomfortable. Ethical leaders speak up when it would be easier to stay quiet. That could mean defending a junior Sailor, confronting inappropriate behavior, or ensuring every voice is heard. Chief Storekeeper Pearl Hall understood this instinctively. As one of the first African American women to serve in the Navy's Women Accepted for Volunteer Emergency Service (WAVES) program during World War II, she entered a service not built for her and helped change it from within. Hall's service was a form of quiet resistance. She faced discrimination and doubt from the very institution she volunteered to support but instead of retreating, she held fast to the standard. She showed her peers that professionalism, competence, and dignity belonged to everyone. Today, Sailors inherit the culture she helped shape, and the responsibility to continue her legacy. When you speak up for someone who's marginalized or silenced, you do more than correct a moment, you reinforce a command culture rooted in trust and moral strength. Courage isn't always loud. Sometimes, it's found in a quiet voice that says, "This isn't right," and a steady leader who refuses to look the other way.

Take the Helm
- **Speak Up for Others** – Step in when junior Sailors are disrespected, ignored, or treated unfairly. Your voice may be the one they don't yet have.

- **See What Others Miss** – Pay attention to subtle signs, who isn't being heard, who's being overlooked, who's carrying invisible weight. Recognize quiet injustices and act before silence becomes acceptance.

Eyes on the Horizon
Leadership means looking out for more than yourself. Speak for those who can't and you'll earn a team that always speaks with you.

Day 105 – Integrity Is a Daily Decision

Ego is the Enemy – Ryan Holiday

"Every day you must decide: will you be your best, or will you give in to ego, pride, and comfort?"

Integrity is a daily discipline and not a one-time declaration. Ryan Holiday reminds us that ego urges us to seek the easy win, to dodge discomfort, and to preserve image. But real leadership begins when we resist that impulse and choose what's right. That choice is especially hard when it comes without recognition, when the right path costs influence, convenience, or popularity. MCPON Delbert D. Black proved how powerful that daily choice can be. Taking office in 1967, Black led during the height of the Vietnam War, a time when the Navy was undergoing massive transformation, racial integration, shifting retention policies, widespread morale concerns, and the professionalization of the enlisted force. With no precedent to follow, he visited commands across the fleet, not to dictate, but to listen. He wore the same uniform as the deckplate Sailor, walked the same spaces, and spoke plainly about standards, responsibility, and pride in service. That daily, visible integrity became his leadership signature. Black modeled change. And in doing so, he became a steadying force in uncertain times because he made the hard right choice again and again. That's the true weight of integrity: choosing it especially when it's difficult

Take the Helm

- **Choose Integrity Over Ease** – When faced with the temptation to cut corners, protect image, or look the other way, pause and choose the hard right. Your example sets the ethical tone for your command.

- **Celebrate Integrity, Not Optics** – Praise Sailors who self-correct or report discrepancies before inspections. Celebrate integrity over image.

Eyes on the Horizon

Integrity doesn't happen by accident. You live it through deliberate discipline. When you choose to do what's right every day, you earn something no rank can give: trust.

Day 106 – Ethics Over Advancement

The 5 Levels of Leadership – John C. Maxwell

"Leaders become great not because of their power, but because of their ability to empower others."

John C. Maxwell reminds us that the most impactful leaders lead with integrity and service, not with status. Advancement in the Navy matters but not more than your values. The strongest leaders don't manipulate evaluations, undercut their peers, or pursue credit at the expense of others. Trust is earned through fairness, results, and a clear commitment to the group's success. Chief Special Warfare Operator Britt Slabinski understood this deeply. Long before he received the Medal of Honor, he was known in the SEAL community for putting his teammates first. During the 2002 Battle of Takur Ghar in Afghanistan, Slabinski led a rescue attempt under heavy enemy fire to recover a fallen teammate. Even in the face of strategic risk and intense scrutiny, his actions were focused on doing right by his people, not protecting his own image. He continued to lead by example and mentor others throughout his career. Slabinski's leadership wasn't about pursuing advancement. Ethical influence has nothing to do with chasing promotion and has everything to do with consistent, values-driven leadership. When you choose character over competition, your impact lasts far beyond your advancement.

Take the Helm
- **Advance With Integrity** – Support your Sailors' success even when it doesn't directly benefit you. Their advancement is a reflection of your leadership.

- **Empower Without Ego** – Avoid competition rooted in ego. Focus on setting the example and letting your influence speak for itself.

Eyes on the Horizon
Your reputation will outlast your rank. Put integrity ahead of ambition and you'll lead with a respect that title alone can't secure.

Day 107 – Never Look the Other Way

The Mission, the Men, and Me – Pete Blaber

"Common sense is not common practice unless leaders demand it."

Pete Blaber teaches that true leaders handle problems head-on, addressing them before they grow. The temptation to ignore toxic comments, procedural violations, or signs of poor morale may feel easier in the moment but every time a leader looks the other way, the standard quietly erodes. Silence becomes endorsement. Over time, those moments shape the command climate more than any policy ever will. MCPON Joe Campa understood this and made it central to his leadership philosophy. He believed that leading from the deckplates meant owning the responsibility, and he challenged Chiefs across the fleet to stop tolerating behavior that undermined good order and trust. Campa frequently warned against letting rank, tradition, or peer loyalty become an excuse for inaction. He expected leaders to walk the spaces, listen to Sailors, and act when something was off. He once said, "If you walk by a problem without fixing it, you've just set a new standard." The mindset accomplished more than enforcing accountability. People gained the courage to pass that expectation forward. Campa's legacy reminds us that command climate isn't built by policies. Command climate is forged by what leaders address and what they allow day after day.

Take the Helm
- **Don't Walk Past Problems** – If you observe unprofessional behavior, whether minor misconduct or serious violations, address it promptly and professionally.

- **Speak Up and Set the Tone** – Teach your team that silence equals approval. Create a climate where all Sailors are responsible for protecting the standard.

Eyes on the Horizon
Ethical failure often begins with small compromises. Look away once, and the damage begins. Stand watch on your command climate. Your silence speaks volumes.

Day 108 – Lead With Conscience, Not Compliance

The Admirals – Walter R. Borneman

"Each man knew that leadership sometimes meant going against the grain of popular opinion or political pressure."

Walter Borneman profiles Admirals Nimitz, King, Leahy, and Halsey, four World War II giants who made decisions not for political favor or convenience, but from a deep internal compass. Each faced moments when compliance would have been safer, but instead, they followed principle. Nimitz, for example, reinforced the Guadalcanal campaign when others urged retreat. He trusted the mission, the Marines on the ground, and the long-term value of a hard-fought stand. That decision changed the course of the war, not because it was popular, but because it was right. King challenged both Navy and civilian leadership when he believed strategic logic demanded it. Leahy advised President Roosevelt with honesty others wouldn't dare voice. Even Halsey tempered instinct with duty when lives were at stake. These leaders showed that real loyalty includes the courage to speak up, take risks, and stand firm in integrity. In the modern Navy, ethical leadership still requires that kind of courage. When policies feel misaligned with purpose, or when pressure tempts you to look the other way, conscience, not compliance, must guide your decisions. When you lead with conviction, your team knows you're standing for something greater than your own comfort.

Take the Helm
- **Speak Truth to Power** – When faced with a decision that challenges your values, ask: "What protects the mission and the Sailors?" Let that guide your action

- **Lead With Integrity** – Be the moral compass in the room. Advocate for what's right, not just what's approved or assumed.

Eyes on the Horizon
Leadership rooted in conscience withstands pressure. The strongest leaders don't follow the crowd. They follow the standard and they take others with them.

Day 109 – Your Example Is Your Influence

The Leader's Bookshelf – Admiral James Stavridis and R. Manning Ancell

"The most effective leaders lead by example by showing up, working hard, listening carefully, and staying humble."

Admiral James Stavridis and R. Manning Ancell highlight a consistent truth: the most influential leaders live the values they expect from others. Their words may set direction, but it's their actions that set the tone. Stavridis often led shoulder to shoulder with his team. He shares a story from his time as a ship captain when he made it a habit to walk the deckplates during the midwatch. One night, long past taps, he sat beside a junior Sailor standing security in the passageway outside the armory. They talked about home, the Sailor's goals, and what it felt like to serve on an unfamiliar ship far from family. The conversation wasn't about performance or protocol but rather about presence. That moment stayed with the Sailor for years and later became the foundation of a mentoring relationship. By showing up unannounced, listening without an agenda, and simply being present, he demonstrated the kind of leader he wanted his team to become. You don't need a speech to lead. You need the discipline to live the values you expect from others. Sailors will follow the standard you set with your actions, not your words. Make sure the message you send is the one you want remembered.

Take the Helm
- **Model the Standard** – Model the behavior you expect. Punctuality, uniform, language, and preparation set the tone long before you give direction.

- **Lead Through Presence** – Use your daily routines as quiet leadership moments. Your Sailors are always watching, even when you're not speaking.

Eyes on the Horizon
Your presence is your platform, so use it wisely. Lead first with action and your words will carry weight when they matter most.

Day 110 – Character Is What Carries You

Resilience – Eric Greitens

"You must never confuse faith that you will prevail in the end with the discipline to confront the most brutal facts of your current reality."

Titles fade. Skills evolve. But character endures. Leadership doesn't truly show itself until you're under pressure facing fatigue, failure, criticism, or conflict. It's in those moments that people decide whether you're worth following. Character is what carries you through and what carries others with you. No one exemplified this more than Admiral James Stockdale. As a Prisoner of War in Vietnam for over seven years, he was tortured repeatedly, held in solitary confinement, and denied basic dignity. But even in that darkness, he led organizing resistance, protecting fellow prisoners, and refusing to betray his values. Stockdale's legacy isn't just one of survival but of preserving honor under the hardest conditions. His influence stemmed not from rank or title but from the power of his example. Stockdale believed leadership forms under pressure, built upon your values. His story shows that character may not ease the path, but it makes leadership worth following.

Take the Helm
- **Stand Firm in Crisis** – When under pressure, return to your values. That's where real leadership begins.

- **Build Character Daily** – Develop your character through daily practice. Self-reflection, consistency, and doing the right thing when it's hard.

Eyes on the Horizon
Character doesn't just shape how you lead. It shapes how you're remembered. Build yours daily, and it will carry you through every test the Navy throws your way.

Day 111 – Hold Yourself to a Higher Standard

The 12 Rules for Life – Jordan B. Peterson

"Set your house in perfect order before you criticize the world."

Jordan B. Peterson emphasizes that before anyone seeks to influence, critique, or lead others, they must take full responsibility for their own life. He uses the 1999 Columbine High School massacre as a stark illustration of this principle. He analyzes the writings of the shooters, whose journals revealed deep nihilism, a desire to destroy, and a belief that the world was so corrupt it deserved to be wiped out. Peterson argues that this mindset stems from a refusal to confront one's own chaos and suffering honestly. Instead of taking steps to fix their immediate surroundings, they externalized blame and sought revenge against society. He uses this tragic example to emphasize that the path to meaning and leadership begins with self-discipline: putting your house in order, however small that house may be. If your logs are sloppy, your uniform is out of regs, or your gear is unsecured, you haven't earned the moral ground to call out the same in others. Peterson's warning is clear: chaos left unchecked in your own world will eventually spill into others. If you want to lead a division, a department, or a watch team with integrity, start by bringing order to your own. Fix your space. Master your craft. Then lead.

Take the Helm
- **Lead From the Front** – When standards shift with qualifications, inspections, or procedures, be the first to adjust, not the last to comply.
- **Make Discipline Visible** – Show up early, stay prepared, and take ownership of every action. Your example sets the pace.

Eyes on the Horizon
If you lead yourself with discipline, your Sailors won't question your expectations. They'll mirror them. Hold yourself to more, and you'll inspire others to do the same.

Day 112 – Let Humility Guide Your Leadership

Grit – Angela Duckworth

"Grit grows when you accept that you are not yet the person you want to become."

Leadership isn't about having all the answers. A willingness to grow is what makes it lasting. Angela Duckworth reminds us that grit begins with the humility to acknowledge you're still becoming the leader you want to be. Rear Admiral Grace Hopper lived this truth every day. Despite revolutionizing Navy computing and helping develop early programming languages, she consistently sought input from junior Sailors and challenged outdated thinking. Hopper's humility didn't lessen her authority. Her credibility only grew because of it. She sat with junior Sailors, listened to their insights, and invited them to question processes. Her grit came not from rigid control, but from her relentless pursuit of improvement and willingness to learn at every level. In her words, "You manage things; you lead people." That distinction matters in every Navy workspace. When you're open about what you don't know and willing to learn from those around you, regardless of rank, you create a command culture where collaboration thrives, and growth is expected. Leadership isn't measured by the loudest voice but by the courage to say, "Teach me."

Take the Helm
- **Ask, Listen, Learn** – Ask for feedback from junior and senior Sailors alike. Humility strengthens your credibility and your effectiveness.

- **Correct Course, Don't Cover Tracks** – When you make a mistake, acknowledge it openly and correct it. Your honesty will reinforce the culture you want to lead.

Eyes on the Horizon
Humility builds bridges that rank can't cross. When your team sees you're still learning, they'll follow your lead and grow with you.

Day 113 – Speak Up When It Matters

Resilience – Eric Greitens

"You don't owe anyone silence when something needs to be said."

Eric Greitens reminds us that leadership is about service, not status. Taking care of your people means protecting their safety, trust, and ability to succeed. That kind of leadership requires the courage to speak up when it matters. True leaders know that raising a concern, even when uncomfortable, is an act of care, not insubordination. The grounding of *USS Guardian* in 2013 was a stark example of what happens when voices go unheard or unspoken. Navigational data showed inconsistencies as the ship approached a protected reef, and some watchstanders reportedly had doubts about the charts but no one voiced those concerns with enough force to halt the evolution. The result was the total loss of a U.S. Navy warship. The problem was not a lack of knowledge. A breakdown in communication and leadership caused the failure. Speaking up might have felt risky in the moment, but silence ultimately cost more: a ship, a reef, and a hard-earned lesson. Leaders protect their teams not just by leading from the front but by ensuring every Sailor feels empowered to speak when something isn't right.

Take the Helm
- **Lead with Courage** – Empower your Sailors to raise concerns during evolutions, briefs, and watches. Make it clear that professionalism includes speaking up.

- **Elevate the Mission** – Challenge procedures or decisions when safety, legality, or mission readiness is at stake. Bring facts, stay respectful, and don't stay silent.

Eyes on the Horizon
Leaders aren't remembered for staying silent. They're remembered for standing up when it counted. One voice can change the outcome. Use yours with courage and clarity.

Day 114 – Protect the Culture You Lead

Legacy – James Kerr

"No one is bigger than the team."

James Kerr explores how the All Blacks, New Zealand's legendary rugby team, built one of the most successful cultures in sports by protecting it relentlessly. He illustrates how culture isn't formed by slogans or hype but by habits and values modeled daily. One of the most vivid examples comes from the All Blacks' "sweep the sheds" tradition. After each game, even world-class players like Richie McCaw would stay behind to clean the locker room. This wasn't symbolic but rather a reminder that humility and accountability apply to everyone, no matter how senior. Kerr details how the team deliberately removed players, even stars, who didn't live up to the culture's expectations. Leadership wasn't about power, but about service to something greater than oneself. Senior players mentored younger teammates, captains took time to connect off the field, and every member had a role in upholding the environment. For naval leaders, the lesson is clear: culture is built or broken by what you allow to pass unchallenged. Whether in a division, wardroom, or command, your daily example either reinforces your values or erodes them. What you walk past, you accept. What you model, you multiply.

Take the Helm
- **Set the Climate** – Conduct regular climate check-ins and act on the feedback. Your responsiveness shows you care about more than just metrics.

- **Defend the Culture** – Address toxic behavior early. Don't let poor attitudes or favoritism infect the culture you've worked to build.

Eyes on the Horizon
Culture defines your legacy as a leader. Guard it with the same urgency you give the mission, because both are inseparable.

Day 115 – Make the Ethical Call, Not the Easy One

The 33 Strategies of War – Robert Greene

"What matters is not just winning, but how you win."

It's easy to rationalize shortcuts in the name of success, but Robert Greene reminds us that victory without integrity is hollow and often short-lived. A leader who wins through deception or unethical shortcuts may achieve results in the moment, but they leave behind a trail of disillusionment and fractured cohesion. This can manifest as Sailors losing faith in the chain of command, cutting corners themselves, or disengaging from the mission. Greene makes this clear by highlighting Queen Elizabeth I's refusal to pre-emptively strike Catholic factions within her realm despite pressure to do so. She resisted the easy path of purging dissent in favor of a patient, strategic defense of her legitimacy building alliances and securing long-term stability. Her restraint proved that ethical decisions can be stronger than force and more enduring than fear. For naval leaders, ethical calls often mean slowing an evolution to ensure safety, enforcing standards even when a friend falls short, or challenging unlawful orders. The easy path may deliver speed or temporary approval, but the ethical one builds credibility and command culture that lasts. How you lead determines whether your influence endures or unravels.

Take the Helm
- **Protect What Matters Most** – When faced with a choice between what's easy and what's right, pause and ask: "What protects my team, my integrity, and the Navy's trust?"

- **Reinforce Ethical Wins** – Model tough ethical decisions during training scenarios. Prepare your team to act with moral clarity under pressure.

Eyes on the Horizon
Winning at any cost is a hollow pursuit. The true measure of a leader is not just in achieving results, but in how those results are earned through integrity, restraint, and unwavering ethical standards that strengthen the team long after the moment has passed.

Day 116 – Trust Is Earned in the Small Things

With the Old Breed – Eugene B. Sledge

"You had to earn the respect of the men around you, not by words but by daily deeds."

In war, speeches don't build trust, actions do. Eugene B. Sledge's memoir of the Pacific campaign shows that the strongest bonds among Marines weren't forged in battle, but in the small, daily acts of consistency: cleaning a weapon without being told, standing a full watch in miserable conditions, or sharing scarce supplies. Leadership in *With the Old Breed* wasn't loud or boastful. Steady actions and daily reliability left the lasting impression. That lesson still holds today. Trust is earned not through dramatic displays, but through the discipline to do your job the right way every single time. Sledge respected the leaders who carried their weight without complaint and enforced standards without ego. In the mud of Peleliu and Okinawa, there was no room for boasting, only competence and character. Those who showed up with humility and upheld the mission in the hardest moments earned unshakable trust. That kind of leadership requires integrity in the small things. Because when the big moments come, your team will follow the person who proved day after day that they could be counted on.

Take the Helm
- **Earn It Daily** – Execute daily leadership tasks, briefs, and reports with precision and care. These small actions shape how your Sailors see you.

- **Do the Little Things Right** – Acknowledge and correct minor errors early. Your attention to detail will show your team they can rely on you when it counts.

Eyes on the Horizon
Big moments reveal leaders, but small actions build them. When you show consistency, humility, and quiet strength each day, you earn a trust that no rank or speech can command.

Day 117 – Your Influence Shapes the Next Leader

The War of Art – Steven Pressfield

"You are a role model whether you choose to be or not. You influence others every day by your choices."

Steven Pressfield shows that real leadership works behind the scenes, steadily shaping those nearby without fanfare. He describes resistance as the invisible force pulling us toward laziness, fear, and comfort. Leaders who push through that resistance, not for credit, but for the good of the team, become the quiet force behind someone else's growth. Your example sets the tone for watchstanders, qualification boards, and command climate. Whether you realize it or not, your discipline becomes someone else's standard. Few leaders embodied this more than Admiral Arleigh Burke who led from the front and later institutionalized leadership development as CNO, shaping how generations of officers learned to lead. He believed your example mattered more than your words. Decades later, the Navy reinforced that principle through CPO 365. Every Chief is reminded that leadership is not about position but about preparing others to succeed after you. Leadership is legacy and when you lead with intention, consistency, and care, your influence lasts long after you leave the space. You don't just complete the mission, but instead you build the next generation.

Take the Helm

- **Lead Beyond Your Tour** – What you teach and how you carry yourself will shape your Sailors' leadership long after you transfer.

- **Model What You Want Repeated** – Your actions teach more than your words. Lead like someone is learning from you because they are.

Eyes on the Horizon
Your example becomes someone else's foundation. Lead with integrity today, and your leadership will echo far beyond your watch.

Day 118 – Finish Strong, No Matter the Billet or Tour

Can't Hurt Me – David Goggins

"When you're done, you're only 40% done."

David Goggins teaches that our minds quit long before our bodies do and that greatness lies beyond the point where most people stop. Goggins pushes through physical pain, mental fatigue, and overwhelming odds, not because it's easy, but because finishing strong is a mindset. Whether it was SEAL training, ultra-endurance races, or operating as a Navy Chief, Goggins lived by a principle that applies directly to naval leadership: how you finish matters more than how you start. When others ease off near the finish line, ethical leaders lean in harder. Goggins never let position or timeline dictate his intensity. In his final days of military service, instead of slowing down, he doubled down, training harder, mentoring more, and setting an uncompromising example for younger Sailors. Goggins knew that legacy rests less on past achievements and more on how you carry yourself through your final watch. In the Navy, your final eval, turnover binder, or mentoring session may shape how others remember you and what standard they carry forward. Coasting at the end may feel justified, but it leaves a gap. Finishing strong doesn't just close out your tour. The way you finish sets the tone for how to carry the torch forward.

Take the Helm
- **Lean in at the End** – Whether it's your last month on sea duty or the final watch of deployment. Show up early, give full effort, and lead with pride.

- **Set the Lasting Example** – Encourage Sailors rotating out to leave something better than they found it. Standardize turnover, train replacements, and finish like a professional.

Eyes on the Horizon
How you finish is how you'll be remembered. End every chapter with effort, pride, and the example you want others to follow.

Day 119 – Live the Legacy You Want to Leave

Helmet for My Pillow – Robert Leckie

"The only thing that counted was the character of the man next to you."

Robert Leckie's memoir of combat with the 1st Marine Division in the Pacific offers a raw truth: in war, medals, ranks, and titles fade. What endures is character. The Marines who earned respect weren't always the loudest or the most decorated, but those who stood their ground, shared hardship, and looked out for their fellow man without hesitation. In the foxholes and firestorms, what mattered most wasn't how many medals someone wore, but instead whether they could be counted on to carry their weight and stand by their brothers when it mattered most. Leckie revered the Marines who lifted others, shared the last rations, took the tougher post, and reminded the team they were in it together. That kind of legacy, the kind that outlives your rank, comes from the relationships you build and the character you display in how you treat others. In today's Navy, your influence doesn't just come from your billet or your decisions but rather comes from how well you steward the trust and morale of your team. Camaraderie flourishes when leaders make it a daily priority, not just a hopeful byproduct. Long after the fight, they'll recall less of what you accomplished and more of how you made them feel while fighting alongside you.

Take the Helm
- **Lead Like It Lasts** – Ask yourself weekly: "If I left today, what would they say about my leadership?" Adjust accordingly.

- **Prioritize Character** – Invest in your Sailors' growth more than your own reputation. Your legacy will live through them.

Eyes on the Horizon
You're writing your leadership story with every action. Make it a story worth repeating, one of honor, courage, and lasting influence.

Day 120 – Honor Is the Foundation of Leadership

Make Your Bed – Admiral William H. McRaven

"If you want to change the world, measure people by the size of their hearts, not the size of their flippers."

Admiral William H. McRaven, a retired four-star admiral and Navy SEAL, served as the commander of U.S. Special Operations Command and oversaw the raid that led to the death of Osama bin Laden. He shares stories from SEAL training where honor wasn't about strength or stamina, but about how you treated the person next to you. He describes watching trainees help each other struggle through freezing surf drills and notes how those who led with humility and grit, regardless of background or size, earned the deepest respect. He praises those who took full responsibility for failures in high-risk missions, and leaders who deflected credit to the team when operations succeeded. In McRaven's world, honor means owning your word, carrying out your duty without complaint, and treating even the lowest-ranking Sailor with respect. These lessons resonate far beyond the SEAL teams. In every division, wardroom, and workspace, honor is the glue that binds teams together and the standard that guides ethical leadership. When you lead with honor, your legacy is one of trust, strength, and service to something greater than yourself.

Take the Helm

- **Honor Every Obligation** – Treat every duty whether ceremonial, mundane, or mission-critical as a reflection of your commitment to Navy values.

- **Lead through Daily Discipline** – Remind your team that leadership starts with honoring your word, your uniform, and your people.

Eyes on the Horizon

When your leadership is grounded in honor, your influence endures beyond titles, billets, or time in uniform. Build a legacy worthy of the Sailors who follow in your wake.

PART II:

LEADING THE TEAM

CHAPTER 5:

BUILDING A WINNING TEAM

Day 121 – Leadership Is a Team Sport

Leading With the Heart – Mike Krzyzewski

"A leader is someone who puts the team first, someone who understands that the group's success is more important than individual achievement."

As one of the NCAA's most successful collegiate men's basketball coaches and a two-time Olympic gold medal–winning head coach, Coach K reminds us that great leaders focus on building their teams rather than competing with them. He explains that trust, safety, and shared purpose are what sustain great organizations. Nowhere is that team-first mindset more deeply embedded than aboard the destroyer *USS The Sullivans*, named after the five Sullivan brothers, George, Francis, Joseph, Madison, and Albert, who all enlisted together and died together aboard *USS Juneau*, which sank at sea during the Naval Battle of Guadalcanal in 1942. Their sacrifice became a national symbol of family, loyalty, and service. Today, the crew of *USS The Sullivans* upholds a culture rooted in unity and embraces the ship's motto of "We Stick Together" as a living leadership principle. It's a command where leaders are expected to care for their people, stand shoulder to shoulder in challenge, and prioritize the group over self. Both history and modern leadership remind us that a team-centered approach outlasts charisma or authority. And that it's the team that becomes stronger than any one individual.

Take the Helm

- **Share the Mission** – Use everyday evolutions like working parties, line handling, inspections as opportunities to develop leaders by trusting Sailors with real responsibility.

- **Build Up, Don't Compete** – Coach instead of control. Help your team learn through ownership, feedback, and small leadership wins.

Eyes on the Horizon

Leadership is about your team's growth. When you lead with purpose and trust, your team becomes stronger than you alone ever could.

Day 122 – Build Trust Before You Need It

The 5 Levels of Leadership – John C. Maxwell

"You build trust with others each time you choose integrity over image, truth over convenience, or honor over personal gain."

Trust is something you build long before it's tested, not something you fabricate in a crisis. John C. Maxwell teaches that true leadership influence is earned gradually through consistent, values-driven behavior. He explains that leaders who reach the highest levels do so not by commanding compliance, but by cultivating credibility. That credibility is forged every day on the watchfloor, in how you correct mistakes, communicate under stress, and prioritize your people over your pride. Nowhere was this principle more evident than in the aftermath of the 2000 terrorist attack on *USS Cole*. When the ship was ripped open by a suicide bombing in Yemen, killing 17 Sailors and wounding dozens more, the crew didn't panic. They executed damage control, fought flooding, treated casualties, and stabilized the ship despite their grief. Their actions were heroic, but they weren't spontaneous. That kind of unity and resolve only comes from trust built beforehand through shared hardship, daily training, and leadership that was present and real. They trusted each other long before the casualty struck, and that made all the difference. It's the bond that shifts a crew from barely surviving to overcoming as one.

Take the Helm
- **Train for Unity** – Use damage control drills, watchstanding rotations, and inspections to build trust through teamwork and shared ownership.
- **Build Trust Before the Crisis** – Create a climate of honesty. When your team knows they can rely on you, they'll be ready when it matters most.

Eyes on the Horizon
Trust is tested in crisis, not developed there. Build it every day through your consistency, humility, and presence.

Day 123 – Set Clear Expectations and Enforce Them

The Toyota Way – Jeffrey K. Liker

"People don't fear accountability when expectations are clear and leadership is fair."

Jeffrey K. Liker explains that excellence isn't created by charisma or intensity, but rather created through clear, repeatable standards, enforced with consistency. At car manufacturer Toyota, every process is documented, every role is defined, and every worker is empowered to stop the line if something goes wrong. This approach, central to Toyota's legendary production system, enables innovation. Liker emphasizes that at Toyota, excellence stems from precise expectations and shared discipline, not personality or pressure. The same principle applies in the Navy. Vague guidance leads to misalignment. Clear expectations create confidence, accountability, and unity across the team. Liker highlights how Toyota leaders set a culture where standards weren't optional, and discipline was everyone's responsibility. Leaders were expected to be visible, involved, and precise, not just in setting the bar, but in holding it. Daily follow-through created a culture where correction wasn't feared but valued as part of building trust and improving together. When leaders communicate expectations clearly and enforce them fairly, they build the same culture: one where Sailors know where they stand, what success looks like, and how to speak up when something goes off course. Clear standards lay the groundwork for any real improvement.

Take the Helm
- **Define the Standard** – Set expectations early and revisit them regularly. Your Sailors should know what success looks like in every evolution.

- **Enforce with Discipline** – Address deviations immediately, with consistency and fairness. Standards only matter if you enforce them.

Eyes on the Horizon
Clear expectations create confident teams. When you lead with precision, your Sailors will respond with performance.

Day 124 – Share the Win to Strengthen the Team

The Five Dysfunctions of a Team – Patrick Lencioni

"When there is no sense of collective success, team members retreat into individual goals."

Patrick Lencioni teaches that cohesive teams thrive when everyone feels connected to a shared mission and celebrated in its success. He warns that when leaders focus only on standout individuals or high-visibility units, trust and collaboration erode. Operational victories depend on everyone, from the bridge to the engine rooms. Leaders who recognize collective effort, not just individual achievement, foster unity that lasts beyond any inspection or evolution. The Navy's performance in Operation Desert Storm showcased how integrated planning and fleet coordination achieve decisive results. Carrier aviation, submarines, logistics, combat systems, and communications all operated in sync. Success came from trust between commands, shared information, and cross-platform support. Sailors in supply and engineering were as essential to the outcome as those in the cockpit or CIC. Leadership made a point to acknowledge that. As Lencioni suggests, when people see how their piece of the puzzle fits into the larger picture and feel that success is truly shared, they grasp why their effort matters. When they believe success belongs to everyone, they claim the mission as their own.

Take the Helm
- **Connect the Dots** – Connect your team's work to larger mission goals. Use briefs and updates to show how each success is a shared one.

- **Celebrate Collective Success** – Recognize departments and divisions as units, not just individuals. Foster pride in collective results.

Eyes on the Horizon
When your team shares the win, they'll share the work. Celebrate together, and you'll build a crew that pushes together.

Day 125 – Team Culture Is a Daily Investment

Atomic Habits – James Clear

"Every action you take is a vote for the type of person you wish to become."

As James Clear explains, culture is forged through daily habits and routine, defining how a team stands up under pressure, beyond what's written in briefings or checked in inspections. He emphasizes that lasting trust and performance are rooted in consistent values-driven behavior. This shows up in how leaders run stand-up briefs, handle mistakes, recognize effort, and maintain standards. Culture takes root in the daily realities of Sailors' work, far more than in any all-hands address. During the Guadalcanal campaign in 1942 to 1943, the Navy and Marine Corps operated in extreme conditions with limited resources, constant threat, and staggering losses. What propelled the force ahead was rooted in a culture of endurance and shared duty, forged on each ship, in every foxhole and fighting position. Commanders earned trust through visible leadership, honest communication, and unwavering standards, even in chaos. Sailors and Marines looked to each other for stability, because that was what had been modeled and reinforced long before combat. The lesson remains clear today: if you want your command to respond with unity when it matters most, you must build that cohesion day by day, through deliberate, consistent leadership.

Take the Helm
- **Be Consistent Daily** – Be intentional in your team's daily rhythms. Start with high-standard briefs, training that matters, and walk-throughs with purpose.

- **Model the Culture** – Correct culture violations the moment they surface. Small acts of inaction create larger cracks in team trust.

Eyes on the Horizon
Your culture is your command climate in motion. Shape it daily or it will shape your team without you.

Day 126 – Develop Confidence Through Training

The Mission, the Men, and Me – Pete Blaber

"Confidence is contagious. So is lack of confidence."

Confidence comes from competence that's been tested under stress long before the crisis hits. This principle becomes real on the deckplates: Sailors gain confidence through repetition, realistic training, and honest feedback. The crew of *USS Mason* embodied this in October 2016 when the ship was targeted by multiple anti-ship missiles in the Red Sea. Their response, engaging threats with SM-2 and ESSM missiles in live combat, was calm, coordinated, and lethal. But it wasn't spontaneous. Behind it was a team that had repeatedly drilled for missile engagements under compressed timelines and uncertain threat vectors, repeatedly under pressure, preparing for exactly this kind of missile scenario. The crew trained with simulated "red alerts" at unexpected hours, forced to interpret incomplete data, prioritize rapidly shifting threats, and execute weapons release procedures under watchful scrutiny. By the time real missiles were inbound, *USS Mason's* team were executing muscle memory built through deliberate, high-intensity training. Blaber's philosophy played out on that combat watchfloor. Their confidence wasn't a product of rank or gut feeling, but the result of rigorous training, steady discipline, and a team that knew it could trust itself.

Take the Helm
- **Train Like It's Real** – Prioritize repetition in training, drills, combat systems checks, and casualty responses should reflect real-world stress.

- **Build Thinking Sailors** – During team reviews, highlight what went right and what can improve. Build confidence through honest debriefs, not empty praise.

Eyes on the Horizon
Real confidence comes not from what you wear, but from the work that earns you the right to wear it. Train hard, lead well, and your Sailors will rise when it counts.

Day 127 – Empower With Clarity, Not Just Freedom

Turn the Ship Around! – L. David Marquet

"Don't just give control; give clarity."

Through firsthand experience, L. David Marquet discovered that giving authority without clear intent turns leadership into neglect. His transformation of *USS Santa Fe* began not by relinquishing control entirely, but by creating clarity around every decision. Instead of relying on orders, Marquet developed a culture of intent, training Sailors to declare what they intended to do, and why. But this model only worked because he first ensured that every member of the crew understood the mission, the constraints, and the desired outcomes. Rather than arising from total freedom, initiative grew from clear expectations and disciplined understanding. When Sailors made errors, Marquet didn't blame poor motivation, he traced the issue back to a lack of clarity. That realization led him to walk the deckplates, teach deliberately, and reinforce standards relentlessly. His leadership shifted from giving answers to enabling judgment. Clarity became the foundation for competence, confidence, and control. By investing in his crew's understanding, Marquet unlocked a level of autonomy that made *USS Santa Fe* one of the best-performing submarines in the fleet. Empowerment is a hands-on investment in shared understanding.

Take the Helm
- **Pair Trust With Tools** – Ensure Sailors have the training, context, and confidence to act decisively.
- **Define Intent First** – Teach your team how to lead within boundaries and make sound decisions without constant input.

Eyes on the Horizon
Empowerment without clarity creates confusion. Set the bounds, give the purpose, then let your team lead the how.

Day 128 – Rebuild Trust to Restore the Team

Grit – Angela Duckworth

"Grit is passion and perseverance for long-term goals."

Angela Duckworth teaches that grit is a sustained commitment to long-term progress, especially after failure. When a crew faces failure, whether from an inspection, casualty, or crisis, it's not enough to recover operationally. Leaders must commit to rebuilding trust, cohesion, and belief in the team. That takes daily perseverance, not quick fixes. The crew of *USS Porter* demonstrated this after their 2012 collision in the Strait of Hormuz, when the destroyer struck a supertanker during a high-traffic transit. The incident, attributed to a breakdown in watch team coordination and situational awareness, led to significant damage and the relief of the commanding officer. Morale was low, and confidence, both internally and from the fleet, had been shaken. A new command triad stepped in not with promises, but with presence. They reset expectations by reestablishing standards of watchstanding discipline, professional conduct, and accountability making clear that complacency and poor communication would no longer be tolerated. Rather than issue sweeping changes overnight, the new triad modeled consistent behavior, enforced small corrections, and held every Sailor, regardless of rank, responsible for creating a safer, sharper ship. These clear, enforced expectations sent a message: the command was serious about doing things right, and every Sailor had a role in that recovery. Their reputation was restored because their leaders stayed focused on earning belief again. One day at a time.

Take the Helm
- **Restore Through Consistency** – After a command setback, lead a reset that includes clear communication, reset expectations, and team-centered goals.
- **Focus on Small Wins** – Reinforce team identity through shared effort. Emphasize cohesion in drills, qualifications, and daily operations.

Eyes on the Horizon
Rebuilding trust starts with daily consistency. Show up, speak honestly, and guide your team back to unity. One evolution at a time.

Day 129 – Give Away Ownership to Grow the Team

The Dichotomy of Leadership – Jocko Willink and Leif Babin

"If you don't empower your team with ownership, they will never take full responsibility."

Jocko Willink and Leif Babin teaches that one of the greatest leadership paradoxes is knowing when to step back. He emphasizes that while leaders must set the standard and take ultimate responsibility, they must also trust their team to lead at their level. Growth doesn't come from micromanagement. Growth happens when there's room to think, act, and take ownership of outcomes. This means giving real responsibility, not tasks disguised as leadership, but space to make decisions and learn. During his deployments to Iraq, Willink implemented this firsthand. He empowered junior SEAL leaders to plan and execute their own missions within commander's intent. Rather than micromanaging every detail, he trained them, guided them, and trusted them. Such trust grows through consistent practice, transparent debriefs, and the courage to give and receive honest feedback. When things broke down, leaders doubled down on ownership instead of taking it away. This approach created teams that operated independently but remained aligned. Willink and Babin's lesson for Navy leaders is clear: delegation without support is abandonment, but purposeful empowerment grows confident, capable Sailors who take responsibility like leaders because they've been treated like leaders.

Take the Helm

- **Delegate With Purpose** – Assign meaningful roles that challenge junior Sailors to lead then mentor them through decisions, not just outcomes.

- **Coach, Don't Control** – Let divisional LPOs run key training, briefs, or boards. Guide them but let them own the results.

Eyes on the Horizon

Delegation isn't a break from responsibility but an investment in your people. Give ownership generously, and you'll grow a team that leads beyond you.

Day 130 – Set Clear Roles to Build Team Accountability

Good to Great – Jim Collins

"Great teams are built on clear roles and disciplined people who understand how they contribute to the mission."

As Jim Collins explains, greatness emerges when raw energy is shaped by disciplined systems and consistent standards. He explains that top-performing organizations rely on clearly defined roles and expectations. Where every evolution depends on tight coordination, clarity is non-negotiable. A ship's ability to perform under pressure doesn't come from last-minute heroics, but from a team that knows exactly who is doing what, when, and why. This principle was tested and proven in the Battle of Midway, where U.S. naval forces overcame superior numbers through decisive leadership and role-based execution. Carrier air groups launched in coordinated waves, destroyers screened effectively, and intelligence teams translated data into action because every unit operated within a clearly defined structure. Far from appearing overnight, that clarity grew through years of preparation and the rigorous "Fleet Problems" between the wars, which sharpened roles, doctrine, and integration across the fleet. These exercises built habits of accountability that showed their value when it mattered most. Just as Collins explains, disciplined people in clearly structured roles create teams that execute and not just guess under pressure.

Take the Helm
- **Define the Win** – Define roles and responsibilities explicitly during pre-deployment briefings and daily drills. Ensure every Sailor knows their specific tasks and how they contribute to the mission.

- **Reinforce Roles Repeatedly** – Conduct regular feedback sessions and after-action reviews to fine-tune responsibilities. Address any ambiguities immediately to maintain accountability.

Eyes on the Horizon
Clear expectations empower your team to perform with confidence and precision. When every Sailor understands their role, the entire command becomes a cohesive force ready to overcome any challenge.

Day 131 – Cultivate a Culture of Continuous Improvement

The Leader's Bookshelf – Admiral James Stavridis and R. Manning Ancell

"A winning team never rests on its laurels. It constantly refines and improves every process."

Admiral James Stavridis and R. Manning Ancell emphasize that great commands never stop asking how they can be better. They show us that true excellence comes from reflecting often, learning constantly, and acting deliberately. Leaders who instill this mindset turn every evolution, inspection, and watch rotation into an opportunity to grow. Stavridis often reflected on how lessons from books and personal experience shaped his commands. During his time as Supreme Allied Commander of NATO and earlier as a surface warfare officer, he prioritized open feedback, inclusive after-action reviews, and learning as a team. He once wrote, "The best commands are humble enough to examine themselves honestly and bold enough to change." In his wardrooms, Sailors were encouraged to speak up about what could be improved, not to assign blame, but to sharpen performance. That atmosphere didn't come from policy, but from leadership that modeled humility and curiosity. Stavridis showed that when a command treats learning as a shared responsibility, it creates a team that adapts faster, communicates better, and leads with confidence into the unknown.

Take the Helm
- **Make Improvement a Habit** – Implement regular, structured after-action reviews. Turn every operation into a chance for constructive feedback and actionable improvements.

- **Model Curiosity** – Encourage every Sailor to contribute ideas for improvement and reward suggestions that lead to measurable gains. Make continuous learning part of your team's DNA.

Eyes on the Horizon
A culture of continuous improvement drives innovation and resilience. When your team is committed to refining its processes every day, you build an organization that can adapt, excel, and lead through any challenge.

Day 132 – Build Trust Through Shared Challenge

Leadership in War – Andrew Roberts

"Real leadership reveals itself when shared hardship creates unity and commitment."

Andrew Roberts illustrates that the most respected and effective leaders aren't made in moments of ease but in hardship shared shoulder-to-shoulder with their teams. Whether describing Winston Churchill walking bombed streets during the Blitz or Napoleon sharing cold nights while on campaign with his soldiers, Roberts emphasizes that true trust is earned through proximity to struggle. Leaders who show up when it's hard, not just when it's convenient, inspire deeper loyalty than any title can provide. Roberts recounts how leaders like Dwight D. Eisenhower won over troops not through charisma alone, but by being present at the front lines, listening to concerns, and showing calm under pressure. He stood among his men in stormy weather on the eve of D-Day, not to give speeches but to share the burden. That kind of leadership cements unity. In every warfare community, from shipboard casualty response to grueling deployments, Sailors don't remember the Power Point slides. What they remember is who stood beside them when things got tough. Trust forged in shared challenge becomes the foundation of resilient command culture and the mark of real leadership.

Take the Helm
- **Share the Strain** – Lead from the front during evolutions, drills, and tough days. Build trust by sharing hardship, not avoiding it.

- **Stand in the Arena** – Use high-pressure moments to unify your team. Highlight how overcoming adversity together strengthens cohesion.

Eyes on the Horizon
Tough moments reveal true leadership. When your team sees you stand tall through adversity, they'll follow you anywhere.

Day 133 – Develop Leaders at Every Level

Naval Leadership in the Atlantic World – Richard Harding and Agustín Guimerá

"Naval leadership has always depended on the preparation of subordinates, not just their obedience."

Every Sailor, regardless of rank, can take the time to coach, challenge, and empower a shipmate to lead. This truth was tragically reinforced during the Falklands War, when *HMS Sheffield* was struck by an Exocet missile. While basic damage control training had been conducted, the crew lacked the depth of preparation and practiced leadership needed to respond effectively in crisis. With few junior Sailors empowered or equipped to take initiative, confusion spread and fire containment efforts broke down. The tragedy highlighted a broader failure, not of courage, but of a leadership culture that hadn't invested enough in developing confident decision-makers beyond the chain of command. It exposed what happens when leaders focus on obedience alone, rather than preparing every Sailor to think, act, and lead under pressure. In contrast, U.S. Navy doctrine prioritizes damage control training at every level because survival at sea can hinge on the ability of junior Sailors to lead when chaos strikes. When every Sailor is trained, trusted, and ready to lead in moments that matter, then the team becomes adaptive and ready for anything. The best leaders help talent grow, one conversation, one challenge, and one opportunity at a time.

Take the Helm

- **Coach With Purpose** – Create opportunities for junior Sailors to lead. Assign meaningful responsibilities, then coach them through the challenges.

- **Invest Early** – Promote leadership development as a team priority. Encourage your Sailors to train their reliefs early and often.

Eyes on the Horizon

Your lasting mark comes from those you empowered to lead long after your watch has ended. Build leaders now, and your team will outgrow any challenge.

Day 134 – Set the Tone from Day One

It's Your Ship – D. Michael Abrashoff

"The most important thing a captain can do is create the right climate."

D. Michael Abrashoff emphasizes that leadership is about climate and not so much about command. He learned quickly aboard *USS Benfold* that Sailors don't wait weeks or months to decide what kind of leader you are. They decide on day one, based on how you listen, how you engage, and what you tolerate. The shift in leadership began the moment Abrashoff took the helm, with him writing, "I didn't change the rules. I changed the atmosphere." Instead of hiding behind policy or tradition, Abrashoff made himself present on the mess decks, in work centers, and alongside Sailors on the deckplates. He asked questions and listened without assumption. He didn't rely on formal authority to gain respect. He earned it by showing he cared about the people and the mission. Within months, retention and readiness soared, not because of new rules, but because of a new tone. As Abrashoff demonstrated, your Sailors will mirror the climate you create. If you want them to act with pride, initiative, and ownership, you must model it from the moment you check aboard.

Take the Helm
- **Lead by Example Immediately** – Use your first interaction with a new team, division, or department to define your expectations and reinforce shared ownership.

- **Be Visible, Be Intentional** – Be visible, consistent, and respectful. How you carry yourself sets the tone more than anything you say.

Eyes on the Horizon
Culture begins the moment you show up. Lead with clarity and care from day one and your team will follow the tone you set.

Day 135 – Recognize and Reinforce What Right Looks Like

The Toyota Way – Jeffrey K. Liker

"You get what you measure and what you reward."

Jeffrey K. Liker emphasizes that lasting excellence is shaped by what leaders choose to recognize. He explains that reinforcing the right behaviors through praise, visibility, and example is more powerful than correcting mistakes alone. If you want your team to value precision, discipline, and initiative, you have to highlight those traits when they appear. Recognition, when tied to values, builds pride and ownership. Your Sailors won't guess what matters. They'll know, because you've shown them. This principle was on full display during the Great White Fleet World Tour from 1907 to 1909. Sixteen U.S. battleships circled the globe to demonstrate America's growing naval power but more than a show of force, it was a showcase of professionalism. Every aspect of the fleet's performance: seamanship, gunnery, uniform appearance, even behavior ashore, was reinforced and recognized as a reflection of Navy standards. Crews gained respect for consistent discipline and excellence, made possible by leaders who modeled, guided, and applauded those actions each day. Liker teaches that healthy culture depends on leaders who define what right looks like and reward it in ways that stick with the team.

Take the Helm

- **Celebrate the Standard** – Recognize outstanding performance publicly and specifically. Tie your praise to the team standards you expect.

- **Make Recognition Meaningful** – Encourage peer-to-peer recognition to create a shared investment in team success and accountability.

Eyes on the Horizon

When done well, recognition tells the team exactly what excellence looks like. When you highlight what right looks like, your team will rise to meet it.

Day 136 – Empower the Next Watchstander

Turn the Ship Around! – L. David Marquet

"Leadership is not about telling people what to do. It's about creating the environment where they can do it."

L. David Marquet shares how he transformed *USS Santa Fe* not by tightening control but by releasing it. He rejected the traditional leader-follower model and replaced it with leader-leader thinking, where Sailors closest to the work were trusted to make decisions. Instead of issuing orders, Marquet encouraged his team to state their intent, think critically, and take ownership of their duties. The shift didn't happen overnight but over time, the crew went from passive compliance to active leadership. Marquet demonstrated that true empowerment happens when leaders step alongside, not away. He restructured the qualification process to develop initiative, not just technical knowledge. Junior Sailors were encouraged to lead within their roles, not just wait for approval. The result was a high-performing, highly engaged crew that outpaced expectations and built confidence at every level. In the Navy, this model offers a powerful lesson: when you invest in developing watchstanders as leaders, not just operators, you don't just improve readiness. You create a resilient team capable of thinking, deciding, and leading long after you've turned over the watch.

Take the Helm
- **Develop Leader-Leaders** – Delegate decision-making where appropriate. Let qualified Sailors lead watch teams, brief evolutions, and own their responsibilities.
- **Shift from Orders to Intent** – Instead of telling them what to do, ask what they intend to do and coach from there.

Eyes on the Horizon
True leadership shows in how ready the next Sailor is to carry on the mission without you. Empowerment builds readiness that lasts.

Day 137 – Lead from the Workcenter, Not Just the Podium

Call Sign Chaos – Jim Mattis and Bing West

"You cannot lead from behind your desk."

Leadership is not measured by how visible you are at the podium. True presence happens on the ground. Jim Mattis and Bing West emphasizes that leaders earn trust by showing up where their people work, listening without ego, and learning the reality of daily operations firsthand. This means leaving the desk and walking the spaces, watching drills, checking maintenance, sitting in on briefs, and talking to Sailors where they stand the watch. Mattis modeled this relentlessly throughout his career. As a Marine battalion commander, he memorized the names and backgrounds of every junior Marine. Another well-known example occurred in 2004 during operations in Fallujah. Instead of staying in a command post, Mattis routinely visited front-line units without announcing his arrival. During one such visit, he sat in the dirt with a group of exhausted infantry Marines, listened to their frustrations, and personally followed up with their battalion commander to address equipment delays and clarify mission intent. His unexpected presence and willingness to listen, without fanfare or entourage, reinforced that leadership was about walking the line with your people. In every warfare community, that same truth applies: when you lead from the workcenter, you lead where trust is built, where standards are enforced, and where the mission actually happens.

Take the Helm

- **Be Where It Happens** – Spend time in your team's workspaces. Ask questions, observe workflows, and offer feedback in real time.

- **Lead With Humility** – Lead key evolutions from the deckplates, not the office, your presence reinforces shared purpose and professionalism.

Eyes on the Horizon

Teams follow leaders who show up. When you're present in their world, they'll follow you anywhere in yours.

Day 138 – Reinforce Standards Without Breaking Spirit

Extreme Ownership – Jocko Willink and Leif Babin

"Discipline equals freedom, but discipline must be balanced with understanding."

Jocko Willink and Leif Babin remind us that real leadership focuses less on cracking down and more on building people up. Standards matter. Discipline matters. But leaders must enforce them in a way that strengthens the team, not suffocates it. In the Navy, that means correcting early, clearly, and with respect. Resentment takes root when leaders overcorrect, enforce rules unevenly, or humiliate in public instead of coaching for improvement. Great leaders use discipline to develop their Sailors, not diminish them. Willink often reflected on his time leading SEAL Teams in Iraq, where strict standards were essential but how they were applied made all the difference. When junior leaders made tactical errors, Willink didn't explode in frustration or assign blame, he owned the failure as a leader and used it to teach. "When a leader takes responsibility for problems and solves them, their team grows stronger," he wrote. He showed that discipline works best when it's paired with trust, communication, and humility. Good leadership shifts the focus from finding failure to creating success. Balancing high expectations with real concern for people inspires teams to aim higher than the minimum required.

Take the Helm
- **Correct With Respect** – Correct performance issues early and professionally. Focus on growth, not punishment.

- **Use Discipline to Build** – Pair standards with support. Use every correction to coach, not condemn.

Eyes on the Horizon
Enforcing the standard doesn't require tearing people down. Correct with clarity, lead with purpose, and your team will rise.

Day 139 – Promote Peer Leadership to Strengthen Cohesion

The 7 Habits of Highly Effective People – Stephen R. Covey

"Leadership is a choice, not a position."

Stephen R. Covey teaches that leadership begins with personal responsibility, stepping up not because of a title, but because the situation demands it. He teaches that those with a proactive mindset create momentum by leading from whatever role they have. This mindset transforms teams. When Sailors demonstrate peer leadership by taking initiative, supporting one another, and upholding the standard without being told, it creates cohesion that no policy can enforce. The 1939 sinking of *USS Squalus* tested this principle under extreme pressure. After a catastrophic failure left the submarine trapped on the ocean floor, the surviving crew faced uncertainty and fear in pitch blackness. But inside that compartment, it wasn't just senior leaders who held the team together. Sailors encouraged each other, maintained order, shared oxygen, and followed emergency procedures with calm discipline. Their mutual trust and shared sense of responsibility became their lifeline. Meanwhile, above the surface, the rescue and salvage effort brought together divers, engineers, and Sailors across the Navy who operated beyond formal lines of authority. Peer-to-peer trust and leadership at every level made the impossible possible. Just as Covey suggests, when people choose to lead regardless of rank, the team becomes stronger than its structure.

Take the Helm
- **Empower Leadership at Every Level** – Empower junior leaders to mentor their peers. Reinforce that leadership exists at every level.

- **Build Peer Accountability** – Create a climate where teammates hold each other to high standards with trust and respect.

Eyes on the Horizon
Leadership has little to do with title and everything to do with action. When your team leads together, they become unstoppable.

Day 140 – Train the Way You Want the Team to Fight

Leadership Strategy and Tactics – Jocko Willink

"You don't rise to the occasion. You fall to your level of training."

According to Jocko Willink, steady performance in high-pressure moments comes from preparation long before the test. He explains that elite teams don't magically execute in chaos. They perform because they've trained in chaos. This means realistic drills, disciplined repetition, and clear expectations at every billet. If your team is sloppy or confused in training, they'll be the same in crisis. But if you build habits through consistent, scenario-based reps, they'll know how to move as one when it counts. Willink draws on his experience leading SEAL Teams in Iraq, where complex urban operations demanded instant decision-making and flawless coordination. The real source of success lay in hours of precise drills, clear communication practice, and what he described as daily "leadership reps." Junior operators knew their roles because they'd trained until they didn't have to think. "You want to be predictable," Willink says. "You want to be boring in execution because that's how you win." The same principle applies in every warfare community: if you want your team to execute with calm under fire, you have to train them relentlessly, realistically, and together.

Take the Helm
- **Drill for Reality** – Run evolutions and drills as close to real-world pressure as possible. Build confidence through repetition.

- **Understand Mission Beyond Task** – Ensure each team member understands not just their task but how it connects to the team's success.

Eyes on the Horizon
How your team trains is how your team fights. Practice with intensity, and your Sailors will perform with unity.

Day 141 – Build a Culture of Mentorship, Not Competition

The 5 Levels of Leadership – John C. Maxwell

"Good leaders build strong teams by multiplying strengths, not guarding them."

John C. Maxwell explains that a leader's highest calling is to develop more leaders, expanding their influence by empowering others. That means sharing knowledge, encouraging growth, and building a culture where success is collective, not competitive. Commodore Oliver Hazard Perry embodied this principle during the Battle of Lake Erie in 1813. Anticipating a complex fight against a superior British force, Perry invested time mentoring his junior officers, each commanding smaller, less experienced ships. He emphasized coordination over competition and built trust across commands. When his flagship *Lawrence* was disabled mid-battle, Perry was able to transfer command to the *Niagara* and continue the fight without confusion or breakdown in execution because every subordinate had been trained to act with independence and unity. His officers stayed focused, supported one another, and executed the battle plan as a team, ultimately securing a landmark American victory. Just as Maxwell teaches, Perry didn't build followers. He built leaders. And in doing so, he proved that a culture rooted in mentorship, not rivalry, multiplies strength and turns shared purpose into historic success.

Take the Helm

- **Empower Through Mentorship** – Foster a mentorship-first mindset. Pair experienced Sailors with juniors to build mutual investment and shared success.

- **Celebrate Shared Growth** – Reinforce that one Sailor's win is the team's gain. Eliminate zero-sum thinking from your leadership culture.

Eyes on the Horizon

Strong teams focus less on internal rivalry and more on growing together. Build a crew that lifts each other, and they'll carry the mission farther than you imagined.

Day 142 – Build Confidence Through Cross-Training

Great by Choice – Jim Collins and Morten T. Hansen

"The best-performing teams prepare obsessively for uncertainty."

Jim Collins and Morten T. Hansen reveal that the most resilient teams are not those with the most resources but those that prepare for disruption. They introduce the concept of "productive paranoia," where elite leaders and organizations train for what might go wrong, even when everything is going right. In the Navy, that mindset is realized through cross-training. Sailors who understand their teammates' systems, procedures, and roles build cohesion and confidence. This depth turns small units into adaptable, mission-ready teams that perform under pressure. During Operation Tomodachi in 2011, following Japan's devastating earthquake and tsunami, U.S. Navy ships and aircraft provided urgent humanitarian aid and nuclear disaster support. Victory hinged on Sailors pushing past their training, ready to improvise and shoulder responsibilities that stretched them further beyond their comfort zone. Culinary Specialists offloaded cargo, Aviation Technicians ran medical supply lines, and junior Sailors took on real-time planning roles. What drove those actions was a culture that demands constant readiness and builds trust and shared responsibility among all rates. Like Collins and Hansen suggest, the most effective teams train not just for their job but for the mission. When every second counted, cross-training turned from a smart practice into a life-saving advantage.

Take the Helm

- **Promote Cross-Rate Understanding** – Cross-train team members within and across watch stations to increase flexibility and understanding of the full mission picture.

- **Train for Flexibility** – Use drills and downtime to rotate roles. Build teams that can back each other up under pressure.

Eyes on the Horizon

Your team's strength is in its depth. Train them to understand each other's lanes and they'll drive the mission forward together.

Day 143 – Anchor Team Morale in Daily Purpose

Man's Search for Meaning – Viktor E. Frankl

"He who has a Why to live for can bear almost any How."

Viktor E. Frankl discovered that even in the most inhumane circumstances imaginable, those who endured did so not through strength alone but through purpose. He recounts how prisoners in concentration camps found reasons to keep going by anchoring their suffering to something greater, whether it was love for a family member, unfinished work, or belief in a future contribution. Frankl's lesson is clear: humans can survive almost anything, as long as they believe their struggle matters. Long hours, stressful evolutions, and the grind of daily watchstanding wear down morale, not because the work is hard, but because purpose can feel lost. Frankl wrote, "Life is never made unbearable by circumstances, but only by lack of meaning and purpose." Connecting routine tasks to mission outcomes boosts results and preserves the team's drive. Motivational words matter less than clear reminders that simple daily actions, from inspecting gear, logging rounds to double-checking a checklist, are what keep the ship afloat and the mission on course. Team morale is resilient when the daily purpose is clear.

Take the Helm
- **Connect Tasks to Mission** – Tie daily work to mission outcomes. Use briefs and debriefs to reinforce the *why* behind every task.

- **Speak to the Purpose** – Use your voice as a leader to elevate purpose. Remind Sailors how their efforts shape the bigger picture.

Eyes on the Horizon
A task list tells them what to do. Purpose tells them why it matters. Anchor their efforts in purpose, and their morale will follow.

Day 144 – Coach Your Team Like a Trainer, Not a Referee

Leading With the Heart – Mike Krzyzewski

"A leader may be the most knowledgeable person in the world, but if the players on his team cannot translate that knowledge into action, it means nothing."

Leadership is about developing people who can think, act, and lead on their own, not about hovering with a whistle. As head coach of the Duke University men's basketball team, Coach K recounts how he intentionally pushed players like Christian Laettner and Bobby Hurley to lead not just through performance, but through coaching others. He empowered them to run huddles, give corrections, and guide younger teammates during both practices and games. When mistakes happened, Coach K didn't immediately jump in to correct. Instead, he watched how his leaders responded and only stepped in when teaching was needed. He believed growth happened by equipping players to recognize and solve problems independently. By the time game day arrived, leadership was already distributed, players were executing from shared ownership and not awaiting direction or orders. It's easy to act like a referee, spotting infractions, correcting checklists, and grading performance. But true leadership is measured by how well your Sailors perform when you're not in the room. When everyone knows what right looks like and feels empowered to coach others, then the team becomes self-sustaining, adaptable, and strong.

Take the Helm

- **Lead With Development** – Use real-time coaching during evolutions, not just post-brief critiques, to reinforce expectations while building confidence.

- **Develop Your Sailors as Trainers** – Encourage peer teaching to build leadership skills across the division.

Eyes on the Horizon

Great teams rise because they receive coaching that goes beyond basic correction. Be the trainer your Sailors need to win.

Day 145 – Lead the Culture, Don't Just Manage It

Start With Why – Simon Sinek

"Great leaders create cultures in which people feel safe to speak up and take initiative."

A strong Navy culture doesn't come from orders alone. Leaders create it through everyday actions with how they communicate, support their teams, and show respect. Navy culture is a quiet rhythm of Honor, Courage, and Commitment, carried out in plain view, day after day. You could see that clearly in 1975, during Operation Frequent Wind, the final U.S. evacuation of Saigon. As South Vietnam collapsed, Navy ships and helicopters became the lifeline for thousands of American personnel and Vietnamese refugees. The situation was chaotic. What held it together was not command and control, but rather it was a culture that had been built beforehand, one rooted in service and care. Sailors and Marines rose to the occasion because they understood the deeper reason behind the mission: a commitment to human dignity, loyalty to allies, and the moral responsibility to protect life, especially those who had trusted and supported the United States throughout the Vietnam War. Sailors and Marines were upholding the Navy's core values by safeguarding both American personnel and South Vietnamese refugees. The mission was humanitarian, defined by courage, sacrifice, and a profound sense of duty to others. And when people believe in that kind of purpose, they carry the culture forward without needing to be told.

Take the Helm
- **Lead the Climate** – Ask your Sailors how they feel about the team, then listen, respond, and adjust your leadership accordingly.

- **Inspire the *Why*** – Create belief in the mission so Sailors feel ownership, not just obligation.

Eyes on the Horizon
Culture is not built in the moment. Lead with values every day, so when the test comes, your team doesn't hesitate to rise.

Day 146 – Strengthen the Team by Knowing the People

Ego Is the Enemy – Ryan Holiday

"Humility allows us to listen. And listening builds trust."

Ryan Holiday reminds us that humility, more than authority, is a leader's most powerful tool. He explains that great leaders close the gap between themselves and their teams through genuine connection. Naval teams don't rally behind titles alone. They follow leaders who know their names, understand their strengths, and listen with respect. Trust is built not in grand gestures, but in small, human moments: conversations, questions, and presence. This lesson stretches back to the early days of the Navy. Aboard *USS Constitution*, Captain Isaac Hull and Captain Charles Stewart earned reputations not just for tactical victories but for how they led their crews. Hull famously walked the deck daily, speaking with Sailors, learning their challenges, and sharing credit for victories. Stewart continued that legacy, building cohesion through fairness and approachability. True to form, their leadership put care for others ahead of personal pride. That connection forged crews who fought with skill, believed deeply in their mission, and trusted each other completely. As Holiday reminds us, when you lead without ego, you create a climate where people give more because they know they matter.

Take the Helm
- **Stay Engaged and Grounded** – Conduct regular, informal check-ins with your Sailors. Get to know their goals, challenges, and motivators.

- **Listen to Build Trust** – Show up in their workspaces, ask questions, and listen with intent. Trust begins where presence meets humility.

Eyes on the Horizon
A connected leader builds a connected team. Know your people, and they'll move mountains for the mission.

Day 147 – Consistency Builds Credibility and Teams

Discipline Equals Freedom – Jocko Willink

"People follow consistency more than charisma."

Jocko Willink shows that discipline not only sharpens the individual but also strengthens a leader's influence and consistency. People aren't drawn to leaders simply for charisma. Consistency, integrity, and a calm presence are what inspire lasting followership. In the Navy, where stress and uncertainty are part of the job, your team needs to know what to expect from you. That means showing up on time, enforcing the standard evenly, and responding to pressure with the same composure every time. Trust in the SEAL Teams didn't go to the loudest voice or longest résumé, but to those who stayed calm, clear, and fair when it mattered most. Their consistency created stability, which allowed junior leaders to take initiative and make decisions without fear of being second-guessed or singled out. When your team can count on how you lead, they stop worrying about how you'll react and start focusing on executing the mission. Predictable leadership doesn't mean being soft. Steadiness under pressure is what gives a leader real strength. That steadiness builds trust. And trust builds teams.

Take the Helm

- **Set the Example Daily** – Maintain consistent leadership tone and standards during briefs, inspections, and informal interactions. Predictability builds team trust.

- **Be Steady, Not Sporadic** – Apply the same expectations across your division. Consistency eliminates confusion and favoritism.

Eyes on the Horizon
Charisma may inspire, but consistency sustains. Show up steady, and your team will show up strong.

Day 148 – Culture Is the Real Mission

Leaders Eat Last – Simon Sinek

"The strength of a team is not in its plan, but in its culture."

Simon Sinek teaches that teams succeed not because of policies or systems but because of the trust and values they live by. Culture defines whether Sailors enforce standards, help each other grow, and stay engaged with the mission. Without the right climate, even great leaders struggle to make an impact. When Admiral Zumwalt assumed the role of CNO in 1970, the Navy was struggling with racial tensions, generational disconnects, and a rigid hierarchy that left many Sailors feeling unheard and undervalued. He saw firsthand that a top-heavy, rule-bound culture was driving talented Sailors out of the service. To counter this, he launched his now-famous "Z-Grams," a series of fleet-wide messages that tackled issues head-on: racial and gender integration, grooming standards, education opportunities, and the role of the Chief's Mess in shaping climate. These reforms challenged the identity of the Navy's traditional power structures. Resistance came swiftly, particularly from senior leaders who believed discipline would collapse and tradition was being eroded. But Zumwalt held the line. He believed a more inclusive, responsive culture would strengthen the force. The courage to push through that resistance is the same courage leaders must summon today: the willingness to confront outdated norms and build cultures that serve both the mission and the people.

Take the Helm
- **Lead by Example** – Reinforce values daily through your tone, presence, and decisions. Your actions define the team's culture.
- **Prioritize Climate** – Invest in command rituals that foster unity. Recognition, training, and shared goals build the fabric of a winning team.

Eyes on the Horizon
Culture is what carries your team when plans change and stress rises. Build it with intention, and your team will carry the mission together.

Day 149 – Build a Bench, Not Just a Team

The 5 Levels of Leadership – John C. Maxwell

"Your legacy as a leader lives on in the people you prepare to lead without you."

John C. Maxwell teaches that the highest level of leadership is about developing leaders who succeed after you. He explains that influence endures only when leaders build others to carry the torch. For those wearing the uniform, your orders will end, but your impact doesn't have to end. The best commands don't collapse when a leader departs. Instead, they keep thriving because the next generation has already been empowered. That lesson came to life during the Battle of Lake Champlain in 1814 where Commodore Thomas Macdonough prepared his crew for every contingency before engaging the British. Knowing he couldn't be everywhere at once, he trained his junior officers and gun crews to rotate positions, shift responsibilities, and lead without waiting for orders. When the battle turned chaotic and casualties mounted, those Sailors executed without hesitation keeping the ship in the fight and turning the tide of the war. Macdonough's approach revealed that the strongest leaders invest in others, ensuring the mission endures beyond themselves.

Take the Helm

- **Develop Your Relief** – Start training your replacement well before turnover. Invest in their leadership, not just their qualifications.
- **Create Leadership Depth** – Identify future leaders in your team and give them stretch responsibilities that prepare them to step up.

Eyes on the Horizon

It's not only about how you lead in the moment but how your team leads in your absence. Build depth, and your legacy will carry on in them.

Day 150 – End Every Watch With a Strong Hand-Off

Call Sign Chaos – Jim Mattis and Bing West

"You owe your successor a unit that is better than you found it."

Jim Mattis and Bing West make it clear: true leadership includes how you leave, not just how you lead. He writes that commanders are custodians of their unit's readiness and that stewardship doesn't end until the last signature is signed and the last brief is delivered. Whether you're handing over a watch, a division, or an entire command, a sloppy turnover is a breach of trust. A strong leader leaves a clean deck, clear expectations, and a ready team. Mattis practiced what he preached. When he handed over command of 1st Marine Division, he didn't just check boxes, instead he ensured the incoming commander had every tool needed to win. He walked the terrain with him, shared unfiltered lessons, and made sure the Marines were positioned for continued success. He warned, "Your name is still on it even after you leave." That mindset applies in every Navy space. Your last impression sets the tone for the next leader and defines the culture you leave behind. Lead to the finish line and hand off with pride.

Take the Helm

- **Own the Turnover** – Document key plans, lessons learned, and recommendations before transferring. Help the next leader pick up where you left off.

- **Finish Like You Started** – Take your final weeks seriously. How you leave shapes the climate your Sailors inherit.

Eyes on the Horizon

Great leaders finish strong. Building up a team before you pass the torch guarantees your influence reaches further and lasts longer.

CHAPTER 6:

MASTERING YOUR INFLUENCE

Day 151 – Influence Is Earned, Not Assigned

The 5 Levels of Leadership – John C. Maxwell

"Position is the lowest level of leadership. The only thing you need to be a boss is a title. But to lead people, you have to earn their permission."

John C. Maxwell teaches true influence starts at the second level: Permission, where people follow you because they want to, not because they have to follow. What earns Sailors' trust isn't rank, but leaders who stand beside them, work hard, and truly care. No one embodied this better than Admiral Arleigh Burke. During World War II, he commanded destroyer squadrons in the Pacific, leading from the front during high-risk operations. His call sign, "31-Knot Burke," came not from a speech or his position but from his insistence on pushing his destroyers to maximum speed, faster than many thought possible, to close with the enemy. His crews understood that he would never ask them to do anything he was not willing to lead himself, and regardless of his rank, they trusted his intent. Burke built trust by being visible, humble, and accountable, and his influence carried far beyond his rank. He proved that real leadership is earned through character and consistency and not command alone.

Take the Helm
- **Build Trust Daily** – Conduct daily walkthroughs in your division or workspace. Use those moments to connect, answer questions, and reinforce that leadership is present and accessible.
- **Earn the Right to Lead** – Follow through on commitments you make to your Sailors, whether it's taking an issue up the chain or providing feedback on a qual, your reliability is your influence.

Eyes on the Horizon
Influence is built day by day through showing up and staying consistent. If your Sailors believe in you, they'll follow you further than your rank ever could.

Day 152 – Communication Builds Command Climate

The Culture Code – Daniel Coyle

"Great cultures are built through thousands of small, clear signals sent through words, tone, and behavior that say 'You belong here.'"

In his study of Navy SEAL Team training at Coronado, Coyle observed how instructors, despite their elite stature, communicated with candidates in ways that consistently conveyed belief, purpose, and shared standards. Even during punishing evolutions, instructors maintained clarity, calm tone, and reinforcing language like, "We're not looking for perfect. We're looking for honest effort." These micro-messages shaped an environment where candidates understood what was expected, why it mattered, and that they were part of something bigger than themselves. The result was both compliance and buy-in. The same principle applies across the fleet. Sailors aren't shaped by messages shouted at all-hands or written into policy. They're shaped by how their leaders speak to them during a late watch, a qualification debrief, or a walk-through in the engine room. When a chief calmly coaches after a mistake, or an officer invites junior input with respect, those acts create the emotional tone of the command. Trust is built not by volume, but by tone, repetition, and how we make people feel with our words. Over time, that becomes climate and climate becomes culture.

Take the Helm
- **Set the Emotional Tone** – Use quarters and watch turnovers to reinforce transparency. Explain command decisions when possible and create space for questions that clarify expectations.

- **Prioritize Upward Voice** – Encourage feedback through anonymous surveys or division-level debriefs to identify blind spots and give Sailors a voice in shaping the work environment.

Eyes on the Horizon
Communication is leadership in motion. It either builds trust or erodes it. If you want your Sailors to speak up in crisis, listen to them in calm.

Day 153 – Clear the Fog Before You Give Direction

Leadership Strategy and Tactics – Jocko Willink

"You can't lead a team out of confusion until you name the confusion."

Jocko Willink teaches that when uncertainty clouds a situation, the worst thing a leader can do is push forward blindly. He emphasizes that clarity must come before command. Teams operating in stress, whether in combat or high-tempo operations, can't perform effectively if they're unsure of the mission, their role, or the desired outcome. Confusion breeds hesitation, and hesitation in dynamic environments leads to failure. Effective leaders pause, assess, simplify, and communicate with precision before taking action. Willink draws heavily on his experience as a SEAL Task Unit Commander in Iraq, where chaotic urban combat often meant plans changed by the hour. In one instance, his team was preparing to enter a contested area when intelligence, terrain, and tasking all shifted within minutes. Rather than charge ahead, Willink called a short tactical pause, gathered his team, and simplified the mission to three clear objectives: secure the building, control the rooftop, and set overwatch. By stripping away ambiguity and aligning everyone on what mattered most, the team executed with speed and confidence. That's the essence of leadership in the fog, not ignoring uncertainty but addressing it head-on and lighting the way forward with clarity.

Take the Helm

- **Communicate with Tactical Clarity** – Before issuing tasking, assess whether your team has clarity. Ask questions to surface assumptions, not just nods.

- **Pause and Clarify Before Acting** – When confusion arises, pause to realign. Reset priorities and confirm roles before resuming execution.

Eyes on the Horizon

You can't lead through confusion until you acknowledge it. When you clear the fog first, your team will follow with confidence, not compliance.

Day 154 – Trust Your Voice When It's Hard to Speak

Daring Greatly – Brené Brown

"Vulnerability is not weakness. It's the courage to show up when you can't control the outcome."

Brené Brown reminds us that the moments that matter most in leadership are rarely comfortable. The courage to speak up, especially when the stakes are high, defines not just individual integrity, but command climate and institutional credibility. The Tailhook Symposium in 1991 became the center of a major scandal when dozens of women, primarily Navy and Marine Corps officers, reported being sexually assaulted or harassed by fellow service members during the event. It was a defining moment that tested whether leaders would prioritize the truth or protect the system. Much of the initial response was shaped by silence, defensiveness, and institutional fear, but in the months that followed, it was the voices of junior Officers and leaders across the fleet, those willing to speak candidly about what had happened and what needed to change, that helped push the service toward necessary reform. Brown teaches that vulnerability is about showing up with courage when truth is hard. Those who stepped forward raised the standard. They reminded the Navy that real trust is earned when leaders speak openly, listen intently, and act decisively, even when it's uncomfortable. Leadership is about protecting your people and that begins with using your voice when it matters most.

Take the Helm
- **Choose Truth Over Comfort** – Speak directly and respectfully when communicating difficult truths. Your courage will model the tone for your division.
- **Model Accountability in Every Rank** – Be willing to say, "I got that wrong." The example you set in vulnerability can empower Sailors to speak up and lead with integrity.

Eyes on the Horizon
Leadership means using your voice even when it shakes. Speak with courage, and your Sailors will follow your truth, not just your title.

Day 155 – Clarity Is a Leadership Tool

Start With Why – Simon Sinek

"People don't buy *what* you do, they buy *why* you do it and what you do simply proves what you believe."

Simon Sinek's "Golden Circle" model explains that powerful communication flows from the inside out: start with *why*, then *how*, then *what*. In the Navy, we often reverse that, starting with what needs to get done and assuming everyone understands the rest. But Sailors don't follow orders blindly forever. Sailors want to know the *why* behind the task. When leaders begin with *why*, why this matters, why it protects the team, why it improves readiness, it builds trust and motivation. Then comes *how*, the process or standard to follow. And finally, the *what*, the specific task or directive. When you speak in this order, people engage. When you skip the *why*, they just comply. Whether you're leading quarters, issuing a maintenance brief, or preparing for an inspection, the Golden Circle helps eliminate confusion. If a Sailor knows what to do but not why, they may execute but won't adapt. If they understand the *why*, they'll take ownership. Sinek's strategy shows that clarity should guide teams toward purpose, not merely focus on precision. When leaders communicate with that mindset, teams operate with confidence, consistency, and care.

Take the Helm

- **Start with Why** – Anchor every direction with purpose. Explain how the task supports the mission or team.

- **Build Outward** – Clarify how it will be done, then state exactly what action is expected.

Eyes on the Horizon
Confusion is the enemy of execution. When you communicate with clarity and purpose, your team will act with speed and confidence.

Day 156 – Your Tone Sets the Temperature

Daring Greatly – Brené Brown

"Leaders must either invest a reasonable amount of time attending to fears and feelings, or squander an unreasonable amount of time trying to manage ineffective and unmotivated people."

Brené Brown reminds us that tone is one of a leader's most powerful tools and it often sets the emotional climate for an entire team. Your Sailors watch how you respond under pressure, in conflict, or during uncertainty. Vice Admiral Samuel L. Gravely Jr., the first African American to command a Navy ship and achieve flag rank, understood this instinctively. Gravely often faced skepticism and outright racism, not just from adversaries, but from within the Navy itself. Yet rather than respond with anger or defensiveness, he met resistance with calm professionalism and steady execution. In one instance, when assigned to command a crew unfamiliar with working for a Black commanding officer, Gravely didn't lecture or confront bias directly, instead, he modeled fairness and unwavering courtesy. Over time, his tone and conduct defused tension, won over skeptics, and created a command climate grounded in mutual respect. He understood that emotional leadership was about regulating the temperature of the room with consistency and grace, especially when others expected heat. His composure became a model for generations of leaders who learned that calm presence earns deep trust.

Take the Helm

- **Deliver with Respect** – Pay attention to your delivery during watch turnover, quarters, and especially in moments of correction. Your tone affects how Sailors absorb direction.

- **Coach Composure, Build Credibility** – Coach your junior leaders to lead with composure. How they speak to peers and subordinates will influence team trust and cohesion.

Eyes on the Horizon

Leadership isn't just about what you say, but about how you say it when it matters most. Speak with the tone you want your team to carry into every watch

Day 157 – Connection Is the Foundation of Influence

How to Win Friends and Influence People – Dale Carnegie

"You can make more friends in two months by becoming interested in other people than in two years by trying to get them interested in you."

Dale Carnegie teaches that influence is built on sincere interest in others. Your rank may get attention, but connection earns trust. Leaders who know their Sailors' names, listen with curiosity, and ask about their lives beyond the uniform build relationships that strengthen every part of the team. Carnegie emphasizes that people are motivated not by orders, but by feeling seen and valued. When your Sailors believe you care about them, they become more open to feedback, more willing to contribute, and more likely to follow you in difficult moments. Carnegie's approach isn't complicated, but it takes intentionality. Smile. Ask questions. Remember details. Praise honestly. Criticize gently. These small efforts create a climate where Sailors feel respected, not just used. When leaders focus only on performance, they get compliance. But when they focus on people, they earn commitment. Far from undermining discipline, connection forms the trust that makes real influence possible, especially in demanding settings like the Navy.

Take the Helm
- **Know Your People** – Schedule informal engagements, walkthroughs, coffee chats, or after-watch check-ins, to stay personally connected to your Sailors' experiences.

- **Earn Trust in Every Exchange** – Ask genuine questions during your interactions and listen without interruption. Show them that their voice carries weight.

Eyes on the Horizon
Your words carry further when Sailors feel seen. Connection lays the groundwork for real influence. Leadership falls flat without it.

Day 158 – Listen First, Then Lead

Emotional Intelligence – Daniel Goleman

"Leaders who are good listeners understand the emotional needs of their team and respond in ways that build trust."

Daniel Goleman teaches that emotional intelligence, not IQ, is the strongest predictor of effective leadership. One of its foundational elements is empathy, which begins with listening. The pressure to act quickly and speak with authority can drown out the quiet but essential skill of listening. But Goleman makes clear: when leaders pause to listen first, before reacting, correcting, or deciding, they demonstrate emotional self-awareness and social attunement. These aren't soft skills. They're operational ones. Goleman describes empathetic listening as the ability to perceive not just words, but the emotional undertones behind them. Leaders with this skill can defuse conflict, spot burnout, and respond to morale issues before they escalate. He warns that poor listeners often interrupt, rush to solve, or unconsciously invalidate what others are expressing, eroding trust even while trying to lead. Instead, strong leaders stay present, ask clarifying questions, and reflect what they've heard. This creates what Goleman calls "emotional resonance," a feedback loop where people feel seen and understood. In high-stakes Navy environments, this resonance keeps communication flowing during stress and it builds the psychological safety that teams need to adapt and succeed.

Take the Helm
- **Practice Active Listening** – During counseling or debriefs, allow Sailors to speak fully before offering guidance. Resist the urge to interrupt or problem-solve too early.

- **Lead with Listening** – Train your junior leaders to lead with listening. Model empathy, paraphrasing, and patience in how you handle conflict.

Eyes on the Horizon
Listening is more than a leadership technique, but also a show of respect. When your Sailors feel heard, they become more willing to follow your lead.

Day 159 – Presence Is the Message

The 12 Rules for Life – Jordan B. Peterson

"People watch how you walk into the room before they hear what you say."

Jordan B. Peterson teaches that posture, eye contact, and bearing are powerful indicators of both capability and mental preparedness. Rule 1, "Stand up straight with your shoulders back," is more than a call to good posture. It's a lesson in presence. In both humans and animals, physical presence communicates dominance, stability, and trustworthiness. Your presence sets the tone before you speak a word. Your Sailors are always reading your body language for cues: confidence or fear, attention or distraction, steadiness or volatility. Peterson connects presence to responsibility. He explains that when leaders carry themselves with purpose, others feel safer and more focused. This isn't about arrogance, but about composure under stress. When things go wrong, people instinctively look to the person who seems the least shaken. If that's you, your calm presence becomes the team's anchor. Peterson also warns that if your outward demeanor shows uncertainty or disengagement, your words won't carry weight, no matter how correct they are. Presence is leadership without needing a speech. It shows that you're aware, prepared, and ready to bear the burden of command.

Take the Helm
- **Anchor the Space** – Be deliberate about how you carry yourself during evolutions, incidents, and divisional quarters. Your presence sets the tone for how your team responds.

- **Lead With Your Posture** – Balance visibility with calm. Walk the spaces not just to check work, but to show confidence, connection, and care.

Eyes on the Horizon
Your team takes emotional cues from your presence before they act on your orders. Show up steady, and they'll steady themselves around you.

Day 160 – Your Words Shape the Watchteam

Radical Candor – Kim Scott

"What you say as a leader becomes part of the team's story. Say it like it matters."

Kim Scott teaches that leaders shape culture not just by what they do, but by what they say and how they say it. Talking clearly on watch is not just routine. Strong communication upholds safety and unites the team. The tone, clarity, and precision of your words influence how your team responds under stress. Whether you're conducting a turnover, issuing orders, or running a debrief, your phrasing becomes the team's mental template. If you speak with purpose and precision, your watchteam will follow suit, but if your language is vague or reactive, the watch will mirror that uncertainty. The 1988 downing of Iran Air Flight 655 by *USS Vincennes* is a sobering example of how miscommunication can escalate under pressure. The language used by watchstanders and the CO became increasingly fragmented as confusion grew about the identity and trajectory of an incoming contact. Ambiguous terms, rushed updates, and a lack of shared understanding contributed to a decision that resulted in tragedy. The aftermath led to sweeping reviews of watchstanding procedures and reinforced the Navy's need for standardization, technical clarity, and deliberate communication. These lessons remain relevant today. When leaders choose their words carefully when delivering information, they ultimately shape team behavior, confidence, and outcomes.

Take the Helm
- **Set the Verbal Standard** – Model clear, professional language during briefs and critiques. What you say becomes how your team speaks.

- **Lead with Language** – Coach out unproductive phrases or tones. Help Sailors understand the impact of unclear or careless language.

Eyes on the Horizon
Your language is part of your legacy. Speak with precision and purpose and your team will do the same when it matters most.

Day 161 – Leadership Happens at Every Level

The 7 Habits of Highly Effective People – Stephen R. Covey

"What you do has far greater impact than what you say."

Stephen R. Covey teaches that personal leadership begins with the first habit of "Be Proactive" which means taking responsibility for your choices and influence, regardless of position. This principle shows up in junior Sailors who don't wait for orders to lead, instead they take ownership of their work, speak up in briefs, support their peers, and shape the command climate from the deckplate up. Covey emphasizes that true leadership is about character over control, meaning that people become influential not because of their authority, but because of their example. Leading your peers means stepping up to make a difference, not waiting to be told you can. Naval commands demonstrate this clearly every time senior leadership recognizes junior Sailors as key influencers, those who uphold standards, encourage teamwork, and carry out responsibilities without waiting to be told. These Sailors also embody Covey's second habit, "Begin With The End In Mind." This means setting the tone not just for what needs to be done, but for how to do it with pride and professionalism. By investing in leadership at every level, you build a team that doesn't rely on top-down direction but instead operates on shared ownership.

Take the Helm
- **Elevate Informal Leaders** – Identify and mentor peer influencers within your team. Give them tools to shape behavior and share expectations from within the ranks.

- **Share Leadership Opportunities** – Reinforce that leadership can be earned daily. Recognize informal leaders who model excellence without waiting for a title.

Eyes on the Horizon
Leadership is a mindset. When Sailors lead each other, the team becomes stronger than any one leader could make it.

Day 162 – Team Communication Starts with Shared Language

The Toyota Way – Jeffrey K. Liker

"Without standards, there can be no improvement. But standards must be clear, simple, and understood by everyone involved."

Jeffrey K. Liker emphasizes that standardization is about empowering teams to operate with clarity and confidence. Teams perform best when expectations and communication are shared across every watchbill and evolution. Captain Thomas Truxtun recognized this need in the early U.S. Navy, advocating for formalized signal books to reduce confusion between ships during combat. His efforts laid the groundwork for a common operational language at sea, one that helped improve cohesion during the War of 1812, particularly after the communication breakdowns that plagued early naval engagements. Admiral James Holloway III built on that legacy in the Cold War era, when he led carrier groups operating with a wide range of platforms and allied units. Holloway drove the implementation of standardized air operations communications and task force procedures, ensuring that carrier strike groups could function as cohesive, responsive teams across global operations. Like Liker's example of Toyota's production teams, where every worker follows clearly defined processes and uses shared visual cues to communicate issues on the line, Holloway's Sailors knew exactly what was expected because clarity was built into every operation. Whether in the 19th century with flags or the 20th with encrypted voice comms, the principle remains: shared language builds shared success.

Take the Helm

- **Standardize Communication** – Standardize critical language during evolutions. Use repeatable phrasing for helm, line handling, and casualty responses.

- **Train to Consistency** – Drill communication as a team skill. Evaluate not just what is said, but how clearly and consistently it's delivered.

Eyes on the Horizon

Shared language builds shared performance. When your team speaks the same way, they move as one.

Day 163 – Influence Your Division's Culture From Within

The Five Dysfunctions of a Team – Patrick Lencioni

"If you want a strong team, everyone has to feel responsible for the culture."

Patrick Lencioni teaches that teams falter not from a lack of talent, but from a lack of trust, commitment, and accountability. He emphasizes that culture is shaped by the behaviors that teammates model and tolerate, not just by what leaders mandate. No memo or inspection can reshape division culture on its own. Real change happens when deckplate Sailors uphold standards, confront complacency, and build trust by showing up consistently. Lencioni calls this the "vulnerability-based trust" that allows teams to communicate openly, admit mistakes, and correct each other without ego or fear. When trust is present, healthy conflict becomes productive, peer accountability becomes normal, and results improve, not because someone said so, but because the team has internalized the standard. Culture takes root when informal leaders in the division begin acting with purpose: checking each other's work, offering help without being asked, and holding one another to the expectations they share. Lencioni reminds us that it's this internal ownership, not external enforcement, that transforms a group into a high-performing team.

Take the Helm
- **Model Accountability** – Correct your peers with respect, and invite feedback on your own performance
- **Reinforce Trust** – Be open about mistakes, share credit, and communicate with transparency

Eyes on the Horizon
Culture lives in the conversations Sailors have when leadership isn't in the room. Empower your team to lead each other, and they'll protect the standard when no one's watching.

Day 164 – Speak Up to Make the Team Better

Crucial Conversations – Kerry Patterson, Joseph Grenny, Ron McMillan, Al Switzler

"When the stakes are high, silence is not safe. It's a choice to let problems grow."

Silence, especially in high-stakes situations, is not a neutral act, but a decision to let issues fester. The temptation to stay quiet during a safety concern or team conflict can feel like a way to avoid drama, but it often costs more in the long run. Leaders who build a culture of respectful candor, where Sailors are expected and empowered to speak up, create teams that correct early, operate safer, and trust each other more. The 1986 Space Shuttle Challenger disaster remains one of the most studied examples of the cost of silence. Engineers from Morton Thiokol, the shuttle's contractor, raised serious concerns about the O-rings and the impact of cold weather but their warnings were downplayed and eventually overruled under launch pressure. The failure to listen and to speak up with authority led to the loss of seven lives. For Navy leaders, the lesson is clear: you don't wait until it's safe to speak, instead you speak to make it safe. Courageous communication focuses less on being right and more on safeguarding the mission and your people when it matters most.

Take the Helm
- **Normalize Speaking Up** – Reinforce the expectation that every Sailor has a responsibility to speak up especially when safety or standards are at risk.

- **Train for Tough Conversations** – Practice difficult conversations through scenario training. Give your team tools and language to speak up with professionalism.

Eyes on the Horizon
Silence protects problems. Give your Sailors the voice to speak up and they'll protect the team before you ever have to.

Day 165 – Speak With Confidence to Build Team Trust

Leadership Strategy and Tactics – Jocko Willink

"Uncertainty is where leadership lives. And in uncertainty, your tone must communicate clarity and confidence."

Jocko Willink explains that leadership is often forged in moments when the path forward isn't clear but the leader must be. In SEAL operations, he recounts how teams relied on calm, decisive communication under fire, especially during room clearings and urban combat scenarios in Ramadi. When chaos erupted, the leader's voice was the anchor. Short, clear commands delivered with confidence made the difference between hesitation and synchronized action. In a casualty, tight maneuvering, or high-tempo operations, your tone becomes a leadership tool steering your team before the situation does. Willink recounts an operation where a firefight broke out earlier than expected. His team took cover, waiting for his call. Willink's voice, transmitted through comms with calm authority, relayed instructions that cut through the confusion. The result: the team moved quickly, adapted, and completed the mission. His takeaway was simple: people hear your tone before they process your words. When you speak with controlled urgency, your team mirrors your confidence. And when you hold the line vocally, your presence is felt, whether on a bridge wing, in a combat zone, or at a whiteboard during planning.

Take the Helm

- **Command Through Tone** – Train junior leaders to communicate with calm authority. Run timed drills that simulate pressure without sacrificing tone.

- **Speak Calm Into Chaos** – During real-world uncertainty, stay deliberate in your speech. When your words are steady, your team becomes steady too.

Eyes on the Horizon
In uncertainty, your team listens not just for answers but for confidence. When you speak with clarity, you give them the courage to act.

Day 166 – Your Voice Carries Beyond the Moment

Start With Why – Simon Sinek

"Leaders inspire action not by telling people what to do, but by communicating why it matters."

Simon Sinek teaches that effective leadership communication begins with purpose. At the center of Sinek's Golden Circle is the *why*, the belief that drives everything else. Sailors may follow orders because of rank but they give their full effort when they believe in the reason behind those orders. When leaders communicate the *why* with clarity and conviction, they create ownership, not just compliance. Admiral Elmo Zumwalt understood this better than most. As CNO in the early 1970s, he issued over a hundred "Z-Grams," brief, powerful messages that not only announced policy changes, but explained the rationale behind them. Whether addressing racial integration, grooming standards, or family support, Zumwalt always communicated with purpose. He didn't just change the Navy's policies, he reshaped its culture by speaking to values, not just procedures. His voice, grounded in belief and vision, still echoes in how we lead today. When leaders speak from *why*, their words carry further because they connect to something deeper than tasking. They connect to trust.

Take the Helm
- **Begin With Purpose** – Incorporate "why this matters" into daily briefs, quarters, or training. Connect each task to a bigger picture.

- **Speak With Vision** – Craft your key messages with intention. What you say today can set the tone for weeks of performance.

Eyes on the Horizon
Your words echo long after the brief ends. Lead with clear, purposeful words, and your team will sustain your vision wherever they go.

Day 167 – Public Words Should Match Private Actions

Ego Is the Enemy – Ryan Holiday

"It's not about saying the right thing. It's about being the person who doesn't need to say it."

Ryan Holiday draws deeply from Stoic philosophy, a school of thinking created by ancient Greek thought leaders. This philosophy, rooted in ancient Greece and Rome, teaches that virtue, self-discipline, and inner character matter more than recognition or outward success. It emphasizes controlling one's response to events, living according to principle, and focusing on what's within one's power, especially in the face of adversity. In his examination, Holiday remind us that true leadership earns trust through silent discipline. The Stoics believed that virtue is revealed not through declarations but through action. You can't inspire Sailors with rhetoric if your example doesn't hold up under scrutiny. When leaders act with integrity behind closed doors, their words carry more weight when spoken aloud. Epictetus, a former slave turned Stoic philosopher and teacher in ancient Rome, advised, "Don't explain your philosophy. Embody it." That's the bar for ethical leadership. The moment you say, "This is who we are," your team begins watching for whether you live up to it. Holiday's point is simple: ego wants the credit, but character does the work. Whether you're briefing an inspection or motivating your division, what matters isn't how well you speak but how well you've led when no one was watching.

Take the Helm
- **Align Words with Daily Example** – Use your all-hands messages, divisional talks, and public comments to reinforce behaviors you already model consistently.

- **Reinforce Messages Through Presence and Action** – Follow up motivational talks with walk-throughs, feedback, and visible reinforcement of what you said.

Eyes on the Horizon
Words build momentum but example sustains it. When your team sees that your message isn't just talk, they'll turn belief into execution.

Day 168 – Use the Microphone to Build the Mission

Start With Why – Simon Sinek

"Communication is not about being heard. It's about being believed."

Sailors are constantly bombarded with instructions, schedules, and reports. What cuts through the noise are words that provide clarity and purpose. Leaders must use every opportunity, whether the 1MC, a quick moment on the mess decks, or morning quarters, to reinforce why the mission matters. No story illustrates the cost of failed communication more than the sinking of *USS Indianapolis* in July 1945. After delivering components of the atomic bomb to Tinian Island, *USS Indianapolis* was torpedoed by a Japanese submarine on its return transit. Due to a series of critical communication failures, the ship's absence went unnoticed for over four days. The ship was never reported overdue and no one acted on missing movement reports. Compounding the issue, a distress signal believed to have been sent was dismissed and ignored. As a result, over 300 Sailors died in the initial sinking, and nearly 600 more perished from exposure, dehydration, and shark attacks before rescue arrived making it one of the worst naval disasters in U.S. history. The silence was deafening, but in its wake, the Navy reformed how it tracked ships, verified movement reports, and ensured accountability in communication. Even in tragedy, the lesson was clear: communication must be deliberate, confirmed, and driven by care for those we lead.

Take the Helm
- **Craft Purpose-Driven Messages** – Use the 1MC, quarters, or command correspondence to reinforce not just direction but meaning, values, and vision.

- **Establish a Communication Rhythm** – Keep your messages short, consistent, and tied to what Sailors are doing now. Make it actionable, not abstract.

Eyes on the Horizon
Your voice guides and adjusts, shaping how the team stays balanced and ready. Speak with clarity, and your Sailors will move with confidence.

Day 169 – Repeat the Right Message Until It Sticks

The Toyota Way – Jeffrey K. Liker

"Leaders must practice and reinforce the message constantly. The consistency of purpose is what builds belief."

Jeffrey K. Liker reminds us that it's not enough to say the right thing once. Leaders must repeat the message with clarity and conviction until it becomes part of the team's culture. Sailors are flooded with information and only the most consistent and purpose-driven messages rise above the noise. When leaders align their words with action and reinforce them day after day, those ideas take root. Values like accountability, integrity, and communication only shape behavior when they're repeated with meaning, not just frequency. After the 2017 collisions involving *USS Fitzgerald* and *USS John S. McCain*, the Navy undertook sweeping reforms. A key part of that transformation was re-establishing a culture of competence, ownership, and procedural compliance through repeated, intentional messaging. Leaders across the Surface Warfare community emphasized the same themes: watchstanding fundamentals, navigational proficiency, rest and readiness, and ethical decision-making, at every level of training and leadership engagement. These priorities were repeated in stand-downs, classroom instruction, and certification events. Over time, the message stopped being just a checklist and stood as the measure of how work should be done. Through disciplined repetition and leadership alignment, the Navy began rebuilding the focus and warfighting readiness of its surface force.

Take the Helm
- **Lead With Message Discipline** – Identify two to three core behaviors that define success for your division, and reinforce them during briefs, training, and after-action reviews.

- **Link Actions to Purpose** – Tie feedback, positive and corrective, back to these key expectations so Sailors can internalize what right looks like.

Eyes on the Horizon
What you repeat becomes what your team believes. If you want the message to matter, speak it with consistency and live it with clarity.

Day 170 – Use Your Influence to Elevate the Team

The 7 Habits of Highly Effective People – Stephen R. Covey

"You are a transition figure. What you do can elevate others beyond where they started."

Stephen R. Covey emphasizes that great leaders create environments where others thrive. His third habit, "Put First Things First," challenges us to focus on what truly matters: building relationships, developing others, and creating lasting value. Covey's fourth habit, "Think Win-Win," reframes leadership as a shared journey, where success is multiplied through collaboration rather than competition. These habits remind us that real influence does not rely on control or constant visibility. Intentional effort is what makes a team stronger. Covey's most effective leaders invest in people. Sailors across all ranks should prioritize time to mentor, share knowledge, and support growth, even when it's not required. Those who do this don't lead for recognition. They lead because they understand the mission depends on the strength of those around them. Thinking win-win means helping others succeed without keeping score. When you create space for others to rise, you elevate the entire team. The mark of influence is not how many people you direct, but how many you develop.

Take the Helm
- **Prioritize People Over Ego** – Choose actions that build trust, develop others, and create a culture of shared growth.

- **Create a Win-Win Environment** – Elevate your teammates without comparison. When they succeed, the mission succeeds.

Eyes on the Horizon
The true measure of leadership is not how high you rise, but how many you lift along the way. Use your influence to build others up, and your legacy will outlast your position.

Day 171 – Say the Hard Thing With Respect

Crucial Conversations – Kerry Patterson, Joseph Grenny, Ron McMillan, Al Switzler

"Speak the truth clearly and respectfully, or prepare for confusion, resentment, or both."

The authors of *Crucial Conversations* emphasize that difficult feedback must be direct, but it must also be delivered with respect. Leaders often face situations where the standard must be enforced, mistakes must be addressed, and accountability must be upheld. Preserving dignity and promoting improvement matter more than softening hard truths. Leaders who master this balance create stronger, more unified teams. MCPON John Whittet, the second Master Chief Petty Officer of the Navy, played a critical role in shaping the identity of the Chief's Mess during a period of rapid change in the early 1970s. Whittet made it clear that Chiefs must hold each other accountable, not just in performance, but in conduct, professionalism, and tone. When he traveled to fleet commands, he didn't hesitate to confront poor leadership, but he always did so with humility and purpose. His message was unflinching: the CPO community had to be better, not for prestige but because junior Sailors needed leaders who lived the standard. Whittet's approach was about sitting down, saying what needed to be said, and reinforcing that accountability is a form of care. His conversations were hard, but they earned respect and changed behavior.

Take the Helm
- **Preserve Respect Under Pressure** – Use direct, clear language when addressing underperformance but deliver it one-on-one, with respect and context.

- **Lead Tough Talks With Purpose** – Model how to receive tough feedback yourself. Normalize candor by handling it with humility.

Eyes on the Horizon
Correction without respect kills trust. Speak the truth with care, and your words will sharpen your team instead of cut them down.

Day 172 – Emotional Intelligence Strengthens Your Voice

Emotional Intelligence – Daniel Goleman

"Leaders with emotional intelligence read the room, not just the schedule."

Daniel Goleman defines emotional intelligence as the ability to recognize, understand, and manage both your own emotions and those of others. In naval leadership, this means knowing when to speak, how to speak, and why your words matter in the moment. Good communication with Sailors means giving them the context that makes information matter. Goleman's research shows that leaders who tune in to emotional undercurrents are far more effective at influencing behavior, navigating conflict, and building trust. Their words carry weight not because of rank, but because of resonance. Goleman breaks emotional intelligence into five core components: self-awareness, self-regulation, motivation, empathy, and social skill. When a naval leader lacks these, even the most well-intended message can fall flat or worse, do harm. But when you combine empathy with self-discipline and timing, your communication meets people where they are. A well-timed pause, a shift in tone, or simply listening first can turn routine guidance into a moment of real impact. Goleman's studies show that emotionally intelligent leaders foster stronger team cohesion, lower stress levels, and higher performance because their message doesn't just come from the head, but from the heart.

Take the Helm

- **Read Before You Speak** – Pay attention to morale, tension, and stress levels before delivering major guidance. Adjust your tone and timing to match.

- **Develop Emotional Awareness** – Coach your LPOs and JOs to lead with awareness. Emotions shape how messages are received across the deckplates.

Eyes on the Horizon

More than good communication, leadership needs authentic connection to inspire action. When your words resonate with Sailors' reality, they trust what you say and how you lead.

Day 173 – Communicate Unity Through Consistent Tone

Radical Candor – Kim Scott

"Consistency in tone builds consistency in culture."

Scott emphasizes that effective leadership is rooted not only in what you say, but in how you say it and how reliably you say it across time, pressure, and audience. She defines strong leadership as a balance of "care personally" and "challenge directly," warning that inconsistent tone breeds confusion and fractures culture. Beyond pitch and volume, tone shows how present you are, what you intend, and how consistently you lead in stressful moments. Leaders who communicate with one voice, especially across the command triad, foster alignment and shared expectations. Few naval leaders embodied this better than Captain Isaac Hull during his command of *USS Constitution* in the War of 1812. Facing *HMS Guerriere*, Hull maintained calm, composed communication even as his ship took fire. His steady orders and deliberate pacing were mirrored by his officers, creating an atmosphere of discipline rather than chaos. Hull's tone didn't waver. His steadiness reassured the crew, clarified intent, and kept the ship unified in the fight. After their resounding victory, Hull credited his Sailors and kept his post-battle communication humble and consistent with the tone he had carried throughout. His leadership proved that consistency in voice, especially in moments of intensity, can shape not only outcomes, but culture.

Take the Helm
- **Align Tone with Intent** – Coordinate message tone and content with your peer leaders. Avoid undermining cohesion through off-hand comments or inconsistent standards.

- **Create Cultural Consistency** – Check in with Sailors to see how your guidance is received. Tone may unify or divide depending on how it's heard.

Eyes on the Horizon
Leadership isn't just about what you say, but about what your team hears together. Align your tone with purpose, and you'll lead a crew that stays aligned under pressure.

Day 174 – Shape Culture From the Deckplates

The Five Dysfunctions of a Team – Patrick Lencioni

"Culture is not built by authority. It's built by example."

Patrick Lencioni teaches that high-performing teams are built through trust, accountability, and daily behavior. He outlines the fundamental breakdowns that destroy cohesion. Starting with the absence of trust, it can cascade into fear of conflict, lack of commitment, avoidance of accountability, and inattention to results. These dysfunctions thrive when junior leaders withdraw or model self-preservation instead of ownership and vulnerability. But when deckplate leaders lead with transparency and set the tone through example, they begin to repair culture from the bottom up. This is where junior Sailors have more influence than they often realize.

They are often the most observed and emulated members of the team. Their attitude, accountability, and everyday actions create a baseline that others mirror making them quiet architects of command culture. Lencioni emphasizes that trust begins when team members are willing to be real with one another, not always having the answers, but being willing to ask for help and hold each other accountable. When a division officer calmly works through a tough problem, or a work center supervisor owns a mistake publicly, that moment sends a stronger message than any formal policy. Culture is shaped by those daily micro-decisions by who speaks up, who stays steady, and who puts the team before themselves when it counts.

Take the Helm
- **Model What You Expect** – Empower your junior leaders to shape divisional tone through their example. Coach them to lead calmly, visibly, and consistently.

- **Call In, Not Just Out** – After high-stress events, gather the team to talk through lessons learned. Turn performance into culture through shared reflection.

Eyes on the Horizon
Culture doesn't start with command. Lead the culture at your level and the command will follow.

Day 175 – Say It So It Sticks

Crucial Conversations – Kerry Patterson, Joseph Grenny, Ron McMillan, Al Switzler

"When stakes are high and opinions vary, clarity isn't optional. It's leadership."

The authors of *Crucial Conversations* remind us that when emotions rise and consequences matter, effective communication becomes the heart of leadership. In the Navy, where lives, equipment, and missions are on the line, Sailors can't afford vague instructions or unclear intent. High-stakes conversations, whether during a casualty, during a debrief, or while correcting a peer, require leaders to be direct without being disrespectful and firm without being inflexible. When you speak clearly and listen openly, you reduce uncertainty and increase performance. Strong leaders create conditions where truth can be spoken safely and heard constructively. The book emphasizes the power of "making it safe" by focusing on mutual respect and shared purpose, especially in conflict. This means creating a communication culture where Sailors don't just hear guidance, they understand it, internalize it, and act on it. Saying something in a way people remember is not about being loud. Careful word choice and tone carry the message further. Clarity, empathy, and tone are the tools that help leaders be understood when it matters most.

Take the Helm
- **Use Closed-Loop Communication** – Train your team to use clear, repeatable phrasing during evolutions. Standardize communications across all watch stations.

- **Make It Safe to Speak** – Build trust by fostering respectful, two-way conversations, even in tense situations.

Eyes on the Horizon
Clear words create confident action. When your message is simple and direct, your team can move faster and perform better.

Day 176 – Influence is Contagious, Use It Wisely

How to Win Friends and Influence People – Dale Carnegie

"Setting an example is not the main means of influencing another, it is the only means."

Dale Carnegie teaches that influence isn't just what you project, but what others absorb and replicate. Carnegie tells the story of Charles Schwab visiting a steel mill where productivity was lagging. Without scolding anyone or issuing new orders, Schwab quietly asked a shift leader how many heats or steel batches the previous crew had completed. The man replied "six." Schwab picked up a piece of chalk and wrote a large "6" on the factory floor. When the next crew arrived, saw the number, and realized they were being silently measured, they pushed themselves and surpassed the number by producing seven. Schwab never raised his voice, made a threat, or issued a directive. His influence came entirely from subtle example, positive competition, and quiet leadership. This lesson applies directly to Navy life. A junior Sailor who maintains their gear, treats others with respect, and steps in without being asked can shift the tone of an entire workspace. In the Navy's watch teams, maintenance shops, and berthing areas, culture is often shaped more by informal influencers than by formal ones. That's why every Sailor, regardless of rank, must recognize their daily influence. Leadership begins the moment someone starts modeling your behavior.

Take the Helm

- **Lead Sideways and Downward** – Encourage peer leadership by recognizing when junior Sailors use their voice to raise standards or support their team.

- **Use Influence with Intention** – Challenge yourself and your team to lead by example. Even when it feels small, your actions set the tone.

Eyes on the Horizon
Influence spreads. Use yours to make the team better and others will follow your lead without being told.

Day 177 – Drive Performance Through Message Discipline

The Toyota Way – Jeffrey K. Liker

"Consistency of message enables consistency of behavior."

Jeffrey K. Liker teaches that real operational excellence rests on people working together under clear, shared values, not just on structured systems. One of the core tenets of Toyota's leadership model is "Respect for People" and "Continuous Improvement" (*kaizen*), both of which require message discipline. Beyond giving directions, leaders strengthen expectations through steady, purposeful reinforcement. When the message is clear, consistent, and values-driven, behavior follows. Confusion disappears, accountability rises, and the culture begins to self-correct. Liker explains that at Toyota, leaders at every level use uniform messaging to create cultural alignment. Phrases like "go see for yourself" *(genchi genbutsu)* and "ask why five times" aren't just slogans, they're spoken habits that reinforce decision-making, ownership, and reflection. The Toyota Production System thrives because everyone hears and speaks the same language of excellence. In Navy terms, that means Sailors shouldn't be guessing what matters. Your message should echo in every brief, every debrief, and every plan of the day. When the tone and content of your communication align with the values you want to see, performance follows.

Take the Helm
- **Reinforce the Right Messages Daily** – Coordinate consistent phrases and expectations across your leadership team. Use repetition to reinforce what matters most.

- **Create Alignment Through Repetition** – Align tasking, feedback, and praise with core values and mission focus. What you emphasize is what they'll prioritize.

Eyes on the Horizon
The most trusted messages are consistent, not loud. Repeat the right words with discipline, and your Sailors will follow with confidence.

Day 178 – Use Communication to Build Team Identity

Start With Why – Simon Sinek

"People don't want to join a company. They want to join a cause."

Simon Sinek explains that people are drawn to purpose more than process. He highlights how the Wright brothers succeeded against better-funded competitors because they communicated a clear sense of belief: that powered flight would change the world. Their team didn't work for fame or profit, they worked because they shared a cause. That shared belief became their unifying language, keeping the team aligned through setbacks and sacrifice. Purpose was communicated not through marketing or formal mission statements, but through consistent passion, conviction, and shared focus. Sinek draws a parallel to the U.S. Marine Corps, where identity is forged through every word, ritual, and expectation. From the first day of boot camp, Marines are told not just what to do, but "who they are" and "why they matter." Communication reinforces honor, courage, and commitment at every level, from drill instructors to command leaders. This consistency builds discipline, but more importantly, it builds belief. When teams hear the same message of purpose from every direction, they stop seeing themselves as individuals executing tasks and start seeing themselves as members of something greater. That is how communication creates culture, and culture creates pride.

Take the Helm

- **Speak the Mission With Purpose** – Use team-based language during briefs and check-ins. Reinforce shared standards and values as part of a common identity.

- **Use Language to Reinforce Belonging** – Encourage your division to define who they are. Create a motto, standard, or shared expectation that aligns performance with pride.

Eyes on the Horizon

Sailors don't rally around tasks. Sailors rally around belonging. Use your voice to give your team something to believe in together.

Day 179 – Say Less, Mean More

Discipline Equals Freedom – Jocko Willink

"When you speak, speak with purpose. Say what you mean and mean what you say."

Jocko Willink emphasizes that effective leadership is measured not by how much you speak, but by the weight your words carry. He draws from his experience leading SEAL Task Unit Bruiser, where communication in combat had to be immediate, clear, and unquestioned. There was no room for fluff. Jocko explains that his most effective leadership moments didn't come from giving long-winded speeches, but from saying just enough, with precision and intent, to align his team. A brief statement like "Hold the line" or "Watch the flank" could shift an entire operation. The power came not from volume, but from trust built through consistent follow-through. Inside SEAL platoons, leaders who talk too much or inflate their role are tuned out. Junior leaders who spoke rarely but delivered consistently earned more respect than those who tried to motivate with constant noise. In that environment, a few sharp, disciplined words from a respected teammate could center the entire team. That principle translates directly to the fleet: when Sailors know your words have meaning, they'll lean in every time you speak. Say less and mean it more and your message will carry far beyond the moment.

Take the Helm
- **Speak with Precision and Intent** – Use deliberate, concise communication when issuing guidance. Avoid over-explaining or clouding the message.
- **Let Action Prove Your Message** – Reinforce your leadership principles through repeated phrases that your Sailors can remember and repeat.

Eyes on the Horizon
Your words carry more when you use fewer of them well. Say what matters, back it up, and your voice will shape the team long after you leave the space.

Day 180 – Leadership Language Becomes Culture

The 5 Levels of Leadership – John C. Maxwell

"People do what people see and they repeat what leaders say."

John C. Maxwell emphasizes that leadership is not only modeled through behavior but echoed in the words leaders choose to reinforce. He explains how influence grows from positional authority into something deeper: personal example, consistent communication, and a shared language that others adopt. In Maxwell's hierarchy, Level 1 is built on position: people follow because they have to. But starting at Level 2, influence is earned through relationships and communication, and leaders begin shaping culture not by mandate, but by example and repeated, values-based language. What you say in front of your team becomes what they say when you're not there. That's how culture takes root. Maxwell shares that high-performing teams often use simple, powerful phrases to reinforce identity: "We own our results," or "This is who we are." These aren't buzzwords. They're cultural anchors. He highlights how strong leaders create this alignment by repeatedly communicating principles like trust, responsibility, and service until those values become second nature. When leaders use language that reinforces belief, the team begins to echo it. First in meetings, then in daily operations, and finally in peer-to-peer interaction. Over time, the team stops needing reminders because the language itself becomes a compass. The words you choose as a leader should fill the team with purpose and not just fill the space.

Take the Helm
- **Repeat and Reinforce Daily** – Be intentional with the language you use daily to reinforce standards, values, and behavior
- **Speak the Culture You Want to See** – Encourage your team to adopt and echo leadership language that reflects shared purpose and accountability. Your words shape culture.

Eyes on the Horizon
The culture your Sailors live in starts with the language you lead with. Speak with clarity, speak with values and your words will outlast your watch.

CHAPTER 7:
DECISIONS UNDER PRESSURE

Day 181 – Clarity in Chaos

Man's Search for Meaning – Viktor E. Frankl

"When we are no longer able to change a situation, we are challenged to change ourselves."

In moments of chaos, a leader's ability to stay composed, assess priorities, and make timely decisions becomes the anchor for the entire team. Viktor Frankl shares several specific examples that illustrate clarity amid chaos, perhaps the most powerful being his mental practice of imagining himself giving a lecture on the psychology of the concentration camp while still imprisoned in one. In the midst of extreme suffering at Auschwitz and other camps, he envisioned himself standing in a warm, quiet lecture hall, calmly explaining the psychological responses of prisoners. This act of visualization gave him mental distance from the horror, reasserted a sense of purpose, and allowed him to preserve his identity as a doctor and thinker despite his circumstances. This example shows that Frankl didn't just theorize about meaning, and instead applied it in real time, under the most harrowing conditions imaginable. That's what makes his message resonate so deeply with leaders facing their own high-pressure environments. By mentally rehearsing emergencies, visualizing successful responses, and reinforcing team roles beforehand, Navy leaders can prepare themselves to think clearly when the pressure peaks. Frankl's example reminds us that chaos doesn't have to own the moment.

Take the Helm

- **Impose Calm in the Chaos** – In moments of uncertainty, breathe, stabilize the team, and focus on one clear decision at a time. Confident calm leads to effective action.

- **Train for the Fog** – Train your watch teams to default to prioritized responses in chaos. Use drills to build comfort with uncertainty.

Eyes on the Horizon

In chaos, your team needs your clarity more than just your answers. Your voice, tone, and first decision set the course for everything that follows.

Day 182 – Don't Confuse Speed With Leadership

Team of Teams – General Stanley McChrystal

"Efficiency is doing the thing right. Effectiveness is doing the right thing."

Effective leadership is not defined by how quickly you move, but by how effectively you lead others through movement. General Stanley McChrystal illustrates this by describing how the Joint Special Operations Task Force, initially designed for rapid execution, struggled against Al Qaeda's decentralized tactics in Iraq. Despite advanced technology and elite talent, the task force was hindered by rigid hierarchies and siloed information, creating delays at critical moments. McChrystal responded by shifting the organization's focus from speed to shared understanding. He instituted daily video calls with thousands of participants, fostering real-time communication, transparency, and trust. This allowed leaders across the organization to slow down just enough to ensure alignment before acting. Instead of issuing orders from the top, he provided clear intent, empowering junior leaders to make fast decisions with confidence when needed. As a result, the task force became faster and more adaptable because of distributed leadership grounded in trust. When urgency strikes, leaders must pause to align their teams, clarify expectations, and ensure understanding. In chaos, great leadership means having built the systems and trust that allow your team to respond in sync, with purpose and precision.

Take the Helm

- **Lead with Clarity, Not Just Pace** – In pressure moments, pause to align your team, even 15 seconds of clarity can prevent minutes of confusion.

- **Build Trust Before the Storm** – Invest in relationships and communication before the pressure hits. Your influence will carry more than your orders.

Eyes on the Horizon

Leaders earn trust by being transparent, not simply by moving quickly. When you slow down to align your team, you'll move faster when it matters most.

Day 183 – Trust the Training

The Unforgiving Minute – Craig Mullaney

"In the heat of combat, I didn't think. I reacted. And I reacted the way I had been trained."

Pressure reveals your character in combat and command alike. Craig Mullaney's memoir is a powerful reminder that leadership in crisis is not fueled by raw instinct, but by preparation so ingrained it becomes reflex. He writes, "The battlefield is the most demanding performance imaginable," and no amount of intelligence or talent can substitute for practiced readiness. Mullaney's experience as a platoon leader in Afghanistan showed that composure in life-or-death moments was less about courage and more about the consistency of training that preceded it. When faced with sudden contact or a wounded teammate, his actions weren't improvised. Muscle memory, built through countless drills and rehearsals, took over. Mullaney returns to the theme that disciplined, realistic preparation is what separates leaders who endure from those who falter. He reflects on the relentless repetitions at West Point and Ranger School, not as rituals of hardship, but as crucibles of habit. For Navy leaders, this underscores the value of every fire drill, man overboard response, and watchstander evolution. The purpose isn't perfection, but rather predictability under stress. In the unforgiving minute, you revert to what you've rehearsed and if you've trained well, that will be enough.

Take the Helm

- **Harden Habits Under Pressure** – Push for realism in your training environments. Don't sanitize the chaos, simulate it.

- **Master the Fundamentals** – Revisit key procedural drills until responses become second nature. Your team's confidence comes from repetition.

Eyes on the Horizon

In crisis, there's no time to invent the plan. Fall back on training and your team will stand strong when it matters most.

Day 184 – The Risk Is Yours to Own

The 33 Strategies of War – Robert Greene

"A leader must never outsource risk. To lead is to bear the burden of uncertainty."

True leadership begins when the stakes are high and the outcomes uncertain. Robert Greene asserts that a leader who distances themselves from risk is managing, not leading. In war, as in command, risk cannot be delegated. The weight of every decision, the danger, the consequence, the unknown, all of it rests squarely on the shoulders of the one in charge. Greene explores how great military leaders from Xenophon to Napoleon understood that committing fully, even recklessly at times, galvanized their forces and clarified decision-making under stress. He highlights the example of General William Tecumseh Sherman, who famously positioned himself forward in hostile territory during his March to the Sea, not to micromanage but to demonstrate full ownership of the campaign's risks. Greene explains that when leaders absorb uncertainty and accept consequences upfront, they earn credibility and loyalty that cannot be faked. Whether it's launching a mission in uncertain seas, authorizing a degraded equipment evolution, or pushing a watch section through tough conditions, your Sailors must see that you're owning the full weight of the decision and not merely issuing directives. That's what earns trust. That's what defines command.

Take the Helm

- **Own the Consequences** – Don't push risk decisions downward. Own them as the leader and prepare your team to execute with confidence.

- **Model Moral Courage** – In post-ops reviews, be transparent about what you knew, what you assumed, and how you made the call.

Eyes on the Horizon
With leadership comes risk and accountability. When you carry that weight with honesty and courage, your team will carry you with trust.

Day 185 – Lead the Moment You Arrive

Call Sign Chaos – Jim Mattis and Bing West

"The most important six inches on the battlefield is between your ears."

Leadership begins the moment you step into the arena. Jim Mattis and Bing West make clear that mental readiness is the most decisive factor in battle, not rank or experience alone. Mattis didn't wait for permission to lead, but instead he prepared himself mentally long before taking command. He explains how he studied every piece of doctrine he could get his hands on and rehearsed battlefield decision-making scenarios to sharpen his thinking for when it counted. Before assuming command of the 1st Battalion, 7th Marines, he not only committed to memory the names and strengths of subordinate leaders but also developed immediate plans to establish his presence and intent from day one. As soon as he arrived, he joined patrols, asked direct tactical questions, and quickly identified where the unit needed focus. Mattis made it clear through both presence and mental preparation that he wasn't catching up to the role. The role met a leader who had already done the work. Whether you're a new CO, Chief, or Work Center Supervisor, your Sailors need to see that same readiness. Leadership doesn't wait. The moment you arrive is the moment you are watched. Think fast. Act purposefully. You're already on the field.

Take the Helm
- **Arrive Ready to Act** – Practice mental rehearsals before entering high-responsibility spaces. Imagine the casualty, visualize the decision.
- **Lead with Mental Discipline** – Empower your Sailors to act when you're not there. Build their confidence in responding with authority and calm.

Eyes on the Horizon
When the pressure hits, don't wait to lead. The first voice, first choice, and first tone sets the tempo for the entire team.

Day 186 – Don't Chase Perfect; Decide with Purpose

Great by Choice – Jim Collins and Morten T. Hansen

"The signature of mediocrity is chronic inconsistency. The signature of greatness is disciplined action."

Jim Collins and Morten T. Hansen found that the most successful companies during chaotic, high-risk periods didn't surge ahead through bold innovation or sweeping vision. Success came through relentless discipline, adhering to well-defined performance thresholds, maintaining operational consistency, and executing with precision regardless of external volatility. One standout example is the comparison between Southwest Airlines and its less successful competitors during deregulation. While others fluctuated wildly with aggressive expansion or retrenchment, Southwest stuck to a deliberate growth model, expanding only when the data supported it, regardless of market pressure. This idea is captured in the authors' "20 Mile March" principle: commit to steady progress regardless of conditions, never overreaching in good times or retreating in bad. In naval terms, this is the leader who doesn't stall operations waiting for every variable to align. It's the one who adjusts flight schedules due to weather, reassigns teams mid-mission due to shifting risks, or maintains operational momentum during uncertain deployments. They manage complexity through consistency. Discipline, not reactivity, becomes their competitive edge. In tough moments, leadership focuses less on being perfect and more on guiding the team forward with clear, confident choices. One mile at a time.

Take the Helm
- **Avoid the Perfection Trap** – In dynamic situations, don't wait for perfect information. Apply the 80% rule: act when you have most of the facts and can make a sound, responsible decision. Waiting for 100% may be too late.

- **Prioritize Disciplined Execution** – Teach your team to commit to decisions with discipline, not hesitation. Refine after action, not before it.

Eyes on the Horizon
The perfect decision rarely arrives on time. Make the best call you can and lead the adjustment if needed.

Day 187 – Pressure Reveals the Leader

Ego is the Enemy – Ryan Holiday

"Impressing people is easy. Showing up when it matters is what counts."

True leadership is calm, focused, and present when everything else is on the edge of unraveling. Ryan Holiday warns that ego thrives in comfort but disappears when the stakes are real. He shares the story of General William Tecumseh Sherman, who endured public criticism and personal doubt during the Civil War but remained focused on the broader objective of ending the conflict. Sherman ignored the noise, refused to let ego drive his decisions, and executed the March to the Sea with relentless discipline. Holiday uses examples like this to illustrate how leaders succeed by quieting their ego, staying mission-focused, and maintaining clarity when others spiral. He also examines leaders who failed under pressure, people whose ambition outpaced their discipline, and whose need for credit or control led to breakdowns in trust or performance. In each case, ego created fragility. For Sailors, the lesson is simple: when pressure mounts, real leadership means staying grounded, listening first, and acting on principle rather than pride. In the moment that matters most, people follow those who are steady, not those who are loud. Calm is contagious. And under pressure, it's the quiet leader who holds the line.

Take the Helm

- **Lead Without the Spotlight** – Presence under pressure speaks louder than performance during ease. Let your actions do the talking.

- **Ditch the Ego, Keep the Mission** – Stay focused on solutions, not self-image. Your team needs your steadiness, not your status.

Eyes on the Horizon
When stress spikes, leadership is about presence, not noise. Stay steady, and your team will mirror you.

Day 188 – Return to Purpose, Not Just Procedure

Start With Why – Simon Sinek

"When people understand *why*, they are more likely to follow through with *what*."

Simon Sinek explains that great leaders do more than give instructions or directions. Reminding people why the work matters keeps motivation strong. It's easy for Sailors to disconnect from the deeper mission, but when they remember why their work matters, why the evolution, inspection, or qualification exists, they engage with renewed energy. This leadership mindset was essential during one of the Navy's most significant cultural transitions: the integration of women onto combatant ships in the 1990s. Commands that treated the change as mere policy compliance struggled with tension and resistance, but those that framed the shift through the lens of purpose, fairness, mission readiness, and unity, found success. Leaders who reminded their crews that diversity reflects national values helped Sailors move from resistance to respect, anchoring the change in the Navy's larger why. The commands that thrived were the ones that reconnected every Sailor to the core belief that trust, teamwork, and professionalism come before everything else. The lesson endures: when morale is low or the environment is changing, lead by reminding your team what they're part of and why it matters.

Take the Helm
- **Reconnect to the Why** – Reconnect your team to purpose after stressful events. Remind them what their actions protect or enable.

- **Lead with Meaning, Not Just Metrics** – In critiques or course corrections, explain the *why* behind your decisions to foster ownership, not just compliance.

Eyes on the Horizon
When stress breaks focus, lead them back to purpose. A clear *why* steadies the team far more than another checklist ever could.

Day 189 – Lead the Reset

Resilience – Eric Greitens

"Resilience is about how you recharge, not just how you endure."

Eric Greitens challenges the common belief that toughness means pushing nonstop. Instead, he writes, "We bounce back not because we're unbreakable, but because we've learned how to recover." Whether in combat, crisis, or long operations, leaders have a critical responsibility: to pause, re-center, and steer the team toward recovery and renewed purpose. Admiral William T. Sampson's leadership during and after the Battle of Santiago Bay in 1898 exemplifies this principle. After a long and uncertain blockade of the Spanish fleet in Cuba, U.S. ships endured false alarms, tropical heat, and limited resupply for weeks. When the Spanish squadron finally attempted to escape, Sampson's ships sprang into action, decisively defeating the enemy in a running battle. But the most impactful leadership moment came after the fight. Rather than rush to celebrate or redeploy, Sampson gathered his commanders, reviewed the engagement, and reinforced lessons learned. In this period of recovery, crews were publicly recognized for their performance, while captains exchanged honest critiques of their decisions and ship handling. Gunnery, damage control, and coordination protocols were revised in real time, turning a hard-fought victory into a launch point for fleet-wide improvement. Sampson understood that resilience wasn't just winning the fight but also leading the reflection that followed. The U.S. fleet emerged stronger not because it endured, but because it reset with purpose.

Take the Helm
- **Make Space to Reflect and Refocus** – After a crisis or high-tempo stretch, pause to reset the team. Recognize effort, then clarify the path forward.
- **Lead the Emotional Recovery** – Use post-event tone to shape recovery. Lead with reassurance before critique.

Eyes on the Horizon
After the storm, your team watches your tone. Use the reset to rebuild confidence and carry momentum.

Day 190 – Read the Room Before You Respond

Emotional Intelligence – Daniel Goleman

"Effective leaders tune in before they speak out."

Before you speak, you must first read the room: the tone, the tension, and the emotional state of your team. That awareness allows you to respond with precision and not impulse. This might mean shifting the tone of a debrief, softening a correction in front of peers, or choosing presence over pressure in the heat of operations. During the Battle of Leyte Gulf, Admiral William "Bull" Halsey misjudged not just enemy movements, but the emotional undercurrents shaping his own decision-making. Lured by a Japanese decoy force of carriers, Halsey shifted his fleet north, away from the vulnerable invasion force at Leyte. He failed to fully read the operational "room," ignoring signs that the main Japanese thrust would come through the central Philippines, not from the north. Driven in part by pride and a desire for a decisive carrier engagement, his decision left the San Bernardino Strait unguarded and nearly exposed American landing forces to disaster. The crisis was only averted by the heroic stand of *Taffy 3*. Halsey's action wasn't a failure of tactical knowledge, but a failure to pause, assess the emotional drivers at play, and read the broader strategic atmosphere. The lesson endures: even brilliant commanders must remain attuned to the emotional and operational currents around them or risk leading their team into unnecessary risk.

Take the Helm

- **Pause Before You Act** – During briefs and critiques, scan your team's expressions and energy before launching into your message.

- **Lead with Emotional Precision** – Teach junior leaders that emotional intelligence is part of tactical awareness. Emotion drives execution.

Eyes on the Horizon
A sharp message in the wrong moment dulls its impact. Read the room, then lead it.

Day 191 – Default to Action

Discipline Equals Freedom – Jocko Willink

"When in doubt: move. Take action. Get after it."

Hesitation kills momentum. Jocko Willink makes it clear: when the path forward is unclear, discipline keeps leaders acting instead of stalling. Whether you're navigating through heavy traffic from the bridge or assessing unknown contacts, waiting for a perfect answer often results in missed opportunities or compounded risk. The key is to default to movement, guided by training, anchored in principle, and executed with confidence. The answer is rarely to wait. Move, reassess, and maintain control. Lieutenant Commander John D. Bulkeley embodied this mindset during one of the most daring operations of World War II: the evacuation of General Douglas MacArthur and his staff from the Philippines in 1942. With Japanese forces tightening their grip on the archipelago, Bulkeley, commanding a squadron of underpowered PT boats, was ordered to lead the evacuation from Corregidor to Mindanao, a journey of over 500 miles through enemy-controlled waters. Navigation was uncertain, fuel logistics tenuous, and Japanese aircraft and ships posed constant threat. But Bulkeley didn't wait for clearer skies or safer waters. He moved. He executed the mission with boldness, coordinating evasive routes and timing movements to avoid detection. His ability to act decisively under pressure not only ensured MacArthur's safe evacuation but also sent a clear message about the power of initiative and disciplined command. Action turned a single day's triumph into a moment remembered in history.

Take the Helm
- **Act With Discipline** – Reinforce "default-to-action" mindsets in your Sailors. Teach them what action looks like when time is short.

- **Lead Without Waiting** – Build muscle memory in your teams through fast-start drills and compressed-scenario training.

Eyes on the Horizon
Hesitation is a decision. Lead with deliberate action and you'll teach your Sailors how to step up when it counts.

Day 192 – Your Calm Is Their Confidence

Daring Greatly – Brené Brown

"Anxiety is extremely contagious, but so is calm."

In moments of stress, your emotional tone sets the course. A composed leader helps a team regulate their own stress response, creating space for clear thinking and decisive action. According to Brené Brown, staying calm is a learned skill, strengthened by practicing self-awareness and managing emotions. A leader who models steadiness teaches their team that control is possible even in chaos. That truth was never more evident than in the actions of Commander Howard W. Gilmore, skipper of the submarine *USS Growler* during a patrol in the Pacific in 1943. While surfacing to engage an enemy convoy, *USS Growler* came under sudden fire from a Japanese gunboat. Gilmore was wounded on the bridge, several Sailors were already down, and the ship was taking damage. With seconds to act, Gilmore gave the calmest possible order, three words that became legend: "Take her down." He knew it meant sacrificing himself but saving the crew and the vessel. That clarity, under immense pain and pressure, saved his Sailors. The crew followed his voice, steady, composed, intentional, and carried out the order without hesitation. In that moment, it wasn't just a command decision. That calm under pressure unified the crew and left a timeless standard for courage guided by self-control.

Take the Helm
- **Model Emotional Discipline** – Practice delivering briefs, critiques, and casualty responses with controlled, steady tone.
- **Train Calm Under Pressure** – Coach your junior leaders to use calm as a tool not as a default, but as a signal of leadership maturity.

Eyes on the Horizon
Clear, steady guidance cuts through chaos far better than volume ever could. Stay composed, and your team will find their footing in your voice.

Day 193 – Train Decision-Makers, Not Just Operators

The Toyota Way – Jeffrey K. Liker

"People closest to the work should be empowered to improve the work."

High-performing organizations rely on well-built systems and even better-trained people who keep pushing those systems forward. Jeffrey K. Liker emphasizes that long-term excellence depends on empowering individuals at every level to think critically and make informed decisions. As he explains, "Standardization is the foundation for continuous improvement and employee empowerment." It sets the baseline that enables people to see what's off track, address problems, and step up at critical moments. Liker explains that Toyota's strength came not from top-down control, but from developing problem solvers across the organization. He describes how line workers were expected not only to identify issues but to halt production if needed to protect quality because they were trusted with judgment. This same philosophy applies directly to Navy leadership. It's easy to train operators to follow steps. It takes intention to develop Sailors who can evaluate evolving situations and act decisively within commander's intent. Beyond issuing guidance, leaders teach people to think and decide wisely. That's how you build a team that doesn't wait for orders. They know what right looks like, and they lead when it counts.

Take the Helm

- **Build Thinkers, Not Just Doers** – Create opportunities for your team to explain the purpose during drills. Build habits of thought, not just repetition.

- **Empower at the Deckplates** – Empower junior Sailors to make controlled decisions during evolutions then review their logic, not just the outcome.

Eyes on the Horizon
Operators execute. Leaders think. Train for both and your team will thrive when the pressure is real.

Day 194 – Leadership Is a Combat Skill

Leadership in War – Andrew Roberts

"Leadership under fire is not an abstract trait. It is a battlefield skill, refined through repetition and experience."

Andrew Roberts explores how history's greatest wartime leaders proved that battlefield leadership is earned well before the fight. They trained for it. They rehearsed decisions, built disciplined teams, and developed the mental habits required to impose calm on chaos. Admiral Horatio Nelson's performance at Trafalgar in 1805 remains one of the most studied examples of combat leadership in naval history. Facing a numerically superior combined French and Spanish fleet, he split his fleet into two columns and drove directly at the enemy line to divide and overwhelm them. This bold plan relied entirely on his captains understanding his intent without real-time communication, a necessity in an age when signal flags were limited and battle smoke often obscured vision. Nelson spent months prior to the battle personally mentoring his captains, explaining his tactical thinking, and drilling his expectations. That preparation allowed his captains to operate with initiative and confidence even after Nelson was mortally wounded early in the battle. The result: a decisive victory that shattered Napoleon's naval ambitions. Whether you're a junior Sailor executing a DC drill or a watchstander evaluating a contact report, your calm under stress comes from reps and sets. If you treat each of those moments with discipline and urgency, you'll be ready when the real test comes.

Take the Helm
- **Train Leadership Like a Warfighting Skill** – Repetition builds readiness. Drill scenarios where you must make rapid, decisive calls so your leadership becomes second nature under stress.

- **Lead with Tactical Presence** – In every evolution or casualty response, model calm, clarity, and control. Your presence under pressure sets the tone your team will follow.

Eyes on the Horizon
Leadership in chaos is a combat skill. Train it. Practice it. And when it's tested, execute it.

Day 195 – Stay Decisive When the Plan Fails

The Mission, the Men, and Me – Pete Blaber

"When the plan breaks, don't break with it. Adapt and execute."

Plans are necessary, but they're never sacred. Pete Blaber, drawing from years of experience in Delta Force operations, emphasizes that effective leadership is measured not by how well a plan is written, but by how fluidly a leader adjusts when reality no longer fits the script. This means staying calm when evolutions go sideways, when systems degrade, or when information shifts mid-execution. During the Battle of Surigao Strait in 1944, part of the larger Battle of Leyte Gulf, Admiral Jesse Oldendorf commanded a force of older U.S. battleships, cruisers, and destroyers tasked with blocking a Japanese fleet approaching from the south. The plan was for a coordinated night ambush using radar-directed firepower, a perfect setup on paper. But the fog of war took hold: communications broke down, friendly destroyers struggled to coordinate torpedo runs, and the Japanese force maneuvered unpredictably. Oldendorf and his subordinates didn't cling to the script and instead, adapted in real time, maneuvering ships, delegating firing authority, and trusting their teams to act with initiative. The result was a devastating crossfire trap that annihilated the enemy column. The last time battleships would ever face off in naval history. Their success came not from a rigid plan, but from disciplined improvisation under pressure.

Take the Helm

- **Embrace Friction as Normal** – During drills and evolutions, train "what if" scenarios. Force your team to recover from changing conditions.

- **Recenter on the Mission** – When the plan fails, realign quickly with your core objective, and drive your team forward.

Eyes on the Horizon

Plans are helpful until they aren't. Your ability to lead when the script collapses will define your command climate and your crew's resilience.

Day 196 – Crisis Leadership Leaves a Lasting Mark

Fearless – Eric Blehm

"When others run from the fire, the leader steps into it."

Eric Blehm tells the extraordinary story of Navy SEAL Adam Brown, whose courage and composure under fire left a lasting impact not just because of what he did, but how he did it. During a mission in Afghanistan's Kunar Province, Brown exemplified crisis leadership in its purest form. As his team moved through a hostile valley, Brown was shot through both legs and his right hand, already compromised from a prior injury, was rendered useless. Despite his devastating wounds, he refused to quit. Bleeding and in searing pain, he propped himself behind a rock with his left hand, took control of a squad automatic weapon, and continued suppressive fire to protect his team's retreat. Brown radioed accurate enemy positions and held his ground until reinforcements arrived, never once allowing his injuries to shift his focus from the mission. Blehm highlights that beyond sheer bravery, Brown's legacy came from his calm presence, firm resolve, and his refusal to be defined by pain or disorder. His actions under fire became the model his teammates would carry forward long after the firefight ended. When you lead in crisis, you leave behind more than results. What you demonstrate is a model of courage that others will carry forward.

Take the Helm
- **Lead When It's Hardest** – Remind your team that the hardest moments may become the ones they're remembered for. Help them prepare with confidence, not fear.

- **Make Presence Your Power** – After high-stress events, talk about the leadership you saw, not just the performance.

Eyes on the Horizon
Your actions in crisis become your legacy. Step into the fire, and your Sailors will follow.

Day 197 – Own the Outcomes

Principles – Ray Dalio

"If you're not failing, you're not pushing your limits, and if you're not pushing your limits, you're not maximizing your potential."

Ray Dalio explains that the most effective leaders foster a culture where mistakes are confronted directly, not hidden or excused. He recounts a pivotal moment in 1982 when he publicly predicted that the U.S. economy was headed for a massive depression. He testified before Congress, warned clients, and positioned his firm accordingly. But the depression never came. The economy rebounded, and Dalio's prediction turned out to be completely wrong. The consequences were severe. He lost clients, credibility, and nearly shuttered Bridgewater Associates. Rather than shift blame or retreat in shame, Dalio took complete ownership. He analyzed what went wrong, documented his blind spots, and used that failure to build a new system of principles for decision-making. That public accountability became the foundation for Bridgewater's culture of radical transparency, where mistakes are studied, not hidden, and leadership starts with owning the outcome. Bridgewater eventually recovered and became one of the world's largest and most influential hedge funds, managing over $150 billion in assets at its peak. Maintenance failures, mission delays, or team friction may have multiple causes but accountability starts with the leader. Owning the outcome means setting the tone. Sailors are more likely to own their part when they see you own yours. When leaders respond to failure with reflection, not deflection, they build trust, resilience, and operational readiness.

Take the Helm
- **Lead with Ownership** – In post-mortems and debriefs, start with what you could have done better as a leader, then invite the team to reflect.
- **Model Accountability in Action** – Use public moments of ownership to model accountability. Your example will shape your team's response to failure.

Eyes on the Horizon
When you own the outcome, your team owns the mission. Blame divides. Ownership builds trust.

Day 198 – Prioritize, Then Execute

The Lean Startup – Eric Ries

"The only way to win is to learn faster than anyone else."

In fast-changing environments, clarity is a competitive advantage. Eric Ries emphasizes that effective leaders aren't just those who act, but the ones who quickly identify what matters, discard distractions, and move with focus. He illustrates this through the early missteps of IMVU, the startup he co-founded. At first, the team spent six months building a fully featured product based on assumptions about what customers wanted. But when they finally launched, the response was flat. Users didn't engage, and key features went unused. The breakthrough came when the team changed its approach, shifting to a system of rapid, disciplined decision-making. They began releasing small, testable components or minimum viable products (MVPs) and prioritized learning over perfection. Each release focused on one goal: find out what matters most, then execute. This allowed them to identify real priorities, cut waste, and act with greater speed and confidence. This same mindset applies to leadership in the Navy. When the pressure builds, equipment failure, evolving tasking, and limited manning, it's easy to become reactive and overwhelmed. But the strongest leaders don't try to solve everything. They step back, assess the situation, and focus on the decision that brings the most clarity. That's how you stabilize the chaos and get the team moving again. Prioritize. Then execute. Then re-evaluate.

Take the Helm
- **Focus on the First Domino** – In your own mental drills, practice choosing first action, second action, and what to delegate. Build the habit.

- **Slow Down to Speed Up** – Teach your Sailors to stay calm and look for the priority, not just react to the noise.

Eyes on the Horizon
You can't do it all at once. Prioritize with discipline then execute with purpose.

Day 199 – Reflect, Refocus, Regain Control

The Unforgiving Minute – Craig Mullaney

"Leaders don't run from failure. They stand in the heat and grow from it."

It's action that forges leadership and quiet reflection that refines it. Craig Mullaney reflects on how combat and command both require deep introspection after the pressure fades. He writes, "You can't learn if you don't look back, and you can't lead if you don't learn." Whether the result was success or setback, strong leaders pause, analyze, and grow from the experience. Taking time to reflect shapes maturity and resilience, forming the trust that strong leadership depends on. Mullaney describes how moments of failure, especially his early stumbles as a platoon leader, became turning points not because he avoided them, but because he confronted them head-on. He writes, "I stood in the wreckage because that's where leaders belong when things go wrong." His story reminds Navy leaders that growth doesn't happen by burying mistakes or brushing off high-stress moments. It happens by standing still long enough to ask, "What did I learn?" and "How will I lead better next time?" In fast-paced commands, it's tempting to move on quickly. But leadership deepens when you stop, reflect, and carry the lesson forward.

Take the Helm
- **Face the Moment Fully** – After high-pressure events, lead structured reflection with your team. Ask what happened, what worked, and what needs to change.
- **Debrief Yourself** – Reflect personally. Keep a leadership journal or log to track how you performed and what you've learned.

Eyes on the Horizon
Pressure without reflection builds scars. Pressure with reflection builds wisdom. Lead the pause and own the growth.

Day 200 – Don't Wait for the Perfect Moment

The Obstacle Is the Way – Ryan Holiday

"Waiting for the perfect conditions is a fool's game. The obstacle is the way."

Naval leaders don't wait for smooth seas. They navigate through the storm. Ryan Holiday draws from Stoic philosophy to remind us that the ideal moment rarely arrives. This might mean launching with a partial crew, adapting to degraded gear, or making the call when visibility is low. Leadership isn't about control, but about movement in uncertainty. Marcus Aurelius, Roman emperor and Stoic philosopher, lived this principle daily. His reign was marked by unrelenting crisis: the Antonine Plague devastated the population, and prolonged wars tested the limits of Roman endurance. Yet he didn't wait for stability to lead. He governed through sickness, loss, and chaos, all while journaling reminders to himself that purpose must persist in adversity. "You have power over your mind, not outside events," he wrote. "Realize this, and you will find strength." Marcus reminded himself that waiting for the perfect moment is avoidance, not leadership. Holiday's insight is timeless: the leaders who stand out aren't those who avoid friction, but those who use it to sharpen their decisions. Conditions are never perfect, and missions rarely unfold cleanly. But acting in spite of that, and because of it, is what separates those who lead from those who hesitate.

Take the Helm
- **Turn Friction into Fuel** – Coach your Sailors that conditions are never perfect. What matters is how you lead through what's real.

- **Move Through the Obstacle** – During planning, identify what "must happen" even if things aren't ideal and build the plan around executing that first.

Eyes on the Horizon
There is no perfect time, only prepared leaders. Step forward and make the moment count.

Day 201 – Lead with Limited Information

The Art of War – Sun Tzu

"If you know the enemy and know yourself, you need not fear the result of a hundred battles."

Leaders are rarely afforded perfect clarity. Sun Tzu teaches that victory does not depend on omniscience, but on preparation and the ability to act decisively with what you do know. In naval environments where information is fragmented: fog, cluttered tracks, incomplete diagnostics, or unclear intent, knowing your systems, your people, and your own leadership tendencies allows you to act even in the gray. This was on display during Operation Praying Mantis in 1988, when U.S. naval forces retaliated for the mining of *USS Samuel B. Roberts*. Commanders had limited intelligence on Iranian intentions, dispersed threat contacts, and shifting geopolitical constraints. Yet fleet leaders made clear, measured decisions executing a complex joint strike that destroyed Iranian naval assets and oil platforms without escalating into broader war. Surface action groups and carrier strike elements had to interpret inconsistent signals and trust their training to engage with lethal precision. These leaders didn't wait for perfect intelligence. Disciplined initiative, informed by available intelligence and trust in the team, drove their decisions under pressure. That's the essence of leading in the unknown: being prepared enough to move when the window opens, not when all questions are answered.

Take the Helm

- **Act on Incomplete Information** – Train your teams to act with the information available not to freeze while waiting for a perfect picture.

- **Know Yourself, Know Your Team** – During pre-watch briefs or evolutions, emphasize priorities and intent so your team can adapt even when visibility is limited.

Eyes on the Horizon

Certainty is rare. Judgment is essential. Lead with what you know, and your team will learn how to trust both their data and their instincts.

Day 202 – Don't Overcorrect Under Pressure

Radical Candor – Kim Scott

"Great leaders challenge directly but care personally."

Kim Scott emphasizes that leadership isn't about avoiding tough conversations but about delivering them with clarity and care. She writes, "Radical Candor is the ability to challenge directly and show you care personally at the same time." When things go wrong, leaders often overcorrect, getting too harsh, too distant, or too controlling, but overreaction signals instability. Scott draws from her own leadership experience to show how the strongest teams are built on feedback that's honest and human. She notes, "There's a big difference between being clear and being cruel." The Navy's response to the *USS Stark* incident in 1987 is a powerful example of this principle in action. After the frigate was struck by two Iraqi Exocet missiles, killing 37 Sailors, the pressure on Navy leadership was immense. Public outrage, political scrutiny, and internal grief could have led to sweeping, punitive overreaction, but instead of scapegoating the crew or dramatizing failure, leaders focused on measured accountability. The Navy prioritized systemic improvements, refining threat recognition procedures, engagement protocols, and regional ROEs, while maintaining dignity and cohesion within the fleet. Training was intensified across surface forces, not as punishment, but as reinforcement. In the face of tragedy, Navy leaders modeled radical candor. The message was clear: we will correct behavior, protect our people, and uphold standards without undermining trust.

Take the Helm
- **Lead the Tone, Not the Tension** – Avoid emotional overcorrection. Respond to problems with deliberate tone and structure.

- **Balance Clarity with Care** – Coach junior leaders to hold the line without damaging the relationship. Respect builds resilience.

Eyes on the Horizon
Pressure may tempt extremes. Lead with steady correction. Clarity and care build a team that rebounds stronger.

Day 203 – Prepare Your Team to Think, Not Freeze

Mindset – Carol S. Dweck

"The view you adopt for yourself profoundly affects the way you lead others."

Great leaders prepare their teams to think. Carol S. Dweck contrasts two approaches to challenge: a fixed mindset and a growth mindset. She writes, "In a fixed mindset everything is about the outcome. If you fail, you're a failure. If you succeed, you're a success." But in a growth mindset, success is about learning and adapting. If Sailors are conditioned to fear failure, they will freeze when something unfamiliar happens. But if they are encouraged to think critically, ask questions, and adapt through training, then they'll respond with initiative instead of panic. During the Cuban Missile Crisis in 1962, that adaptability was tested daily as U.S. Navy destroyers enforced a high-risk maritime quarantine against Soviet shipping. With unclear rules of engagement, real-time communication delays, and the risk of triggering nuclear escalation, junior officers and enlisted Sailors often had to make judgment calls with global consequences. When destroyers intercepted Soviet submarines, it wasn't rigid adherence to protocol that preserved peace. Success came from Sailors thinking on their feet, reading ambiguous signals, and balancing restraint with resolve. That level of tactical clarity under pressure was the product of commanders who trained their crews to think, not just execute. Because in moments where there is no script, initiative saves lives and prevents war.

Take the Helm
- **Train for Thought, Not Just Reaction** – Build scenario-based training that forces Sailors to apply judgment not just memorize procedures.

- **Reward Initiative, Not Just Accuracy** – Reinforce confidence in ambiguity. Celebrate when Sailors problem-solve, not just follow orders.

Eyes on the Horizon
Checklists train memory. Leaders train mindsets. When your team knows how to think, they won't freeze when the plan breaks.

Day 204 – The Cost of Indecision

Leadership Strategy and Tactics – Jocko Willink

"Waiting too long to make a decision can be just as dangerous as making the wrong one."

Effective leaders must balance patience with decisiveness. Jocko Willink emphasizes that while deliberate thinking has value, hesitation in critical moments can stall momentum, confuse subordinates, and invite unnecessary risk. In dynamic environments like combat or high-tempo operations, leaders will rarely have all the facts. What matters most is making a timely, well-reasoned decision based on the information available because doing nothing is a choice that often makes things worse. Willink shares examples from his time in combat where hesitation nearly led to mission failure. In one instance, a sniper overwatch team observed suspicious movement in a known hotspot but delayed their report while seeking confirmation. That moment of indecision gave the enemy time to reposition and compromised the element of surprise. Willink underscores that leaders must accept the burden of imperfect decisions and be willing to act with confidence and clarity, even under pressure. By making a call and adjusting as new information develops, leaders maintain control and momentum, both of which are critical in uncertain environments.

Take the Helm
- **Avoid the Freeze** –In tense situations, rely on your training and prioritize safety. Imperfect action is often better than delayed action.

- **Decide with Discipline** – Develop decision-making skills in junior leaders by incorporating pressure-based scenarios into training and critiques.

Eyes on the Horizon
The team can recover from a wrong call but not from no call. Decide with confidence, or risk letting the situation decide for you.

Day 205 – Clarify the Next Move

Good to Great – Jim Collins

"Great leaders simplify complexity into a clear next action."

Jim Collins emphasizes that the most effective leaders are those who distill complexity into clarity. Whether recovering from a casualty or responding to new orders, confusion compounds stress. The leader's role is to simplify, focus, and deliver the next right move. When the future is unclear, clarify the next task. One well-communicated move brings stability and in moments of pressure, that clarity is power. This principle defined one of the most critical decisions of World War II that involved Lieutenant Commander Wade McClusky's course correction at Midway. Leading a group of U.S. Navy dive bombers in search of the Japanese fleet, McClusky arrived over the expected target area and found only empty ocean. With fuel running low and no confirmed intelligence on enemy position, he could have turned back. But instead, McClusky made one clear decision: follow a lone Japanese destroyer racing north. That single move, executed without delay and clearly communicated to his formation, led his bombers directly to the Japanese carrier task force. Minutes later, dive bombers crippled *Kaga*, *Akagi*, and *Sōryū*, turning the tide of the Pacific War. McClusky didn't have a full plan but he had the clarity to define the next move, and that was enough to change history.

Take the Helm

- **Shrink the Problem** – After an unexpected issue, give your team a clear, time-bound task, even a small one, to regain momentum.

- **Lead with Calm Direction** – Build trust by focusing on the next right thing, not the whole plan under pressure.

Eyes on the Horizon
In the fog of uncertainty, one clear move lights the path. When you simplify, your team regains confidence and rhythm.

Day 206 – Delegate Under Fire

The Mission, the Men, and Me – Pete Blaber

"The battlefield is too dynamic for one person to control everything. Trust your team."

In high-pressure situations, the instinct to grip tighter can feel like leadership but it often limits performance. Pete Blaber emphasizes that combat leadership requires empowering others to think and act, especially when conditions are fluid and time is limited. Blaber shares examples from his time in Delta Force where the speed and unpredictability of operations made centralized control impossible. Success came from decentralized execution, delegating decision-making down to those closest to the action. That means your Sailors must know the mission, understand your intent, and be ready to act. When the pressure hits, you lead through trust and that trust is built long before the crisis arrives. Over a tense 10-day period in 2016, Iranian-backed Houthi forces launched multiple anti-ship missiles toward destroyer *USS Mason* and nearby vessels in the Red Sea. With limited warning and rapidly closing threats, combat watchstanders had to make real-time decisions, engaging defensive systems, launching countermeasures, and maneuvering the ship. These weren't hypothetical drills. They were life-or-death responses initiated by delegated authority under pre-established command intent. The ship successfully intercepted multiple threats without waiting for external confirmation or micromanagement from higher echelons. That success was built on preparation, trust, and clear empowerment. A modern textbook example of delegation under fire.

Take the Helm
- **Empower the Front Line** – Assign leadership roles before the pressure hits. Ensure your team is confident acting without constant oversight.
- **Lead from the Top, Not the Center** – During high-stress moments, resist the urge to take over. Empower, trust, and verify.

Eyes on the Horizon
In pressure, your reach is limited. Train leaders across your team so your influence doesn't stop where you aren't.

Day 207 – Control What You Can

Man's Search for Meaning – Viktor E. Frankl

"Everything can be taken from a man but one thing: the last of the human freedoms, to choose one's attitude in any given set of circumstances."

In adversity, your greatest power is how you choose to respond. Viktor E. Frankl, writing from the depths of a concentration camp, teaches that even when we lose control over our circumstances, we never lose the ability to govern our attitude and actions. Struck by a Japanese torpedo, *USS Juneau* sunk during the Battle of Guadalcanal in 1942. Of the 700+ crew, about 100 Sailors initially survived, left stranded in open water with minimal supplies and no immediate hope of rescue. In the days that followed, despite injury and dehydration, junior Sailors stepped into quiet leadership. They organized watch rotations, shared what little food they had, and tried to keep morale afloat in the face of despair. There were no lifeboats, no medics, no chain of command. Just men making the choice, hour by hour, to maintain discipline and care for one another as best they could. Few ultimately survived, but those who did credited the endurance of small acts of leadership: the Sailor who calmed others during the night, the shipmate who passed on his last sip of water. In a situation where everything else had been taken, they chose how to respond, honoring what it means to lead when nothing else is left.

Take the Helm
- **Anchor in Response, Not Circumstance** – Teach your team to shift energy from blame to action. Ask, "What can we still do?"
- **Be the Steady Voice** – Model composure when things fall outside your lane. Focus on your sphere of influence, not your frustration.

Eyes on the Horizon
You can't control everything, but you can lead through anything. Your mindset is the anchor in rough seas.

Day 208 – Communicate Like the Mission Depends on It

Crucial Conversations – Kerry Patterson, Joseph Grenny, Ron McMillan, Al Switzler

"When stakes are high, opinions vary, and emotions run strong. Real leaders talk."

When pressure surges, communication is mission critical, not optional. Leaders either rise to difficult moments through skillful dialogue or retreat into silence, defensiveness, or confusion. Whether you're giving an urgent order, clarifying a misunderstanding, or delivering tough feedback, your tone and timing carry weight. In 1988, *USS Vincennes*, operating in the Persian Gulf, mistook a civilian Airbus airliner, Iran Air Flight 655, for an attacking military aircraft. Amid the high tension of a combat patrol, the ship was flooded with conflicting signals: radar contacts, radio traffic, and urgent threat assessments. But instead of pausing to clarify, verify, and challenge assumptions, the team spiraled into misinterpretation. Watchstanders incorrectly reported the aircraft was descending in an attack profile. Communications were rushed, poorly structured, and based on assumptions rather than confirmed facts. There was no deliberate effort to hold a calm, clarifying exchange and within minutes, the airliner was erroneously shot down. The chain of communication, from radar interpretation to command decision, failed because of fear, urgency, and the inability to slow the conversation when it mattered most. *Vincennes* reminds every leader that communication must deliver clarity and guarantee understanding, especially when tension runs high.

Take the Helm
- **Speak with Precision Under Pressure** – In moments of tension, use calm, direct language to maintain control. Panic spreads fastest through your voice.

- **Create Safe Dialogue in Stress** – Make it safe for your team to hear hard truths and speak up when they see risk.

Eyes on the Horizon
Your team can't follow silence or confusion. Speak clearly, lead intentionally, and they'll stand ready when it counts.

Day 209 – Don't Let Stress Shrink Your Thinking

The 7 Habits of Highly Effective People – Stephen R. Covey

"Between stimulus and response is a space. In that space is our power to choose."

Stephen R. Covey teaches that the most powerful tool a leader has is the ability to pause. He writes, "Our ultimate freedom is the freedom to choose our response." In high-stress moments, average leaders react automatically; however, effective leaders claim that space between stimulus and response to stay in control and lead with intention. That space is where clarity is found, and decisions are elevated. This lesson came into sharp focus during the Battle of the Coral Sea, the first carrier-versus-carrier battle in history. U.S. Navy leaders, operating without real-time communication or satellite intelligence, had to make complex decisions amid a shifting fog of war. Rear Admiral Frank Jack Fletcher and his task force operated with limited visibility, conflicting reports, and enormous stakes. But instead of reacting rashly, they chose to pause and reassess at each critical phase, adjusting flight operations, repositioning carriers, and anticipating Japanese movements based on probability, not panic. Their calm, measured responses helped prevent a direct invasion of Port Moresby and marked the beginning of a turning tide in the Pacific. The outcome wasn't perfect but their ability to think under pressure rather than react emotionally saved ships, shaped future doctrine, and laid the groundwork for the victory at Midway just weeks later.

Take the Helm

- **Create Space to Choose** – Teach your watchstanders and leaders to use deliberate pauses. Space to think is space to lead.

- **Anchor in Values, Not Emotion** – Let your principles, not pressure, guide your decision-making under stress.

Eyes on the Horizon

Stress will close your focus. Leadership reopens it. In the pause, find your clarity.

Day 210 – Finish the Fight

Make Your Bed – Admiral William H. McRaven

"If you want to change the world, never, ever give up."

Leadership begins with a clear start and proves itself by seeing the team through every mile to the finish. Admiral William H. McRaven draws on lessons from SEAL training and combat to highlight the power of perseverance. He writes, "Life is full of difficult moments, but it is how you respond to those moments that will define you." Not every challenge ends in minutes or hours. Some test your stamina, demanding leadership that holds the line, keeps the standard, and doesn't lose heart when fatigue sets in. McRaven recounts brutal training evolutions, where quitting was always one step away, and how those who endured weren't always the strongest, they were the most committed. "You will fail," he says, "you will likely fail often. It will be painful... but if you want to change the world, don't be afraid of the circuses, the long swims, and the obstacles." Real leadership is built in those long hauls, when motivation fades, but discipline and grit carry you forward. It's in those tests that teams prove they can finish the mission and come out stronger on the other side.

Take the Helm
- **Lead Through the Grind with Purpose and Presence** – During prolonged stress, acknowledge the grind then lead through it with purpose, routine, and presence.

- **Normalize Pressure as Part of the Mission** – Teach your team that pressure isn't a roadblock but a normal part of accomplishing the mission.

Eyes on the Horizon
Leadership isn't just about how you start but rather about how you finish. Stay steady. Finish the fight.

CHAPTER 8:

LEADING THROUGH CHANGE

Day 211 – Change Requires Leadership First

Turn the Ship Around! – L. David Marquet

"The people on the ship had not changed. It was the leader who changed."

L. David Marquet shares how transformation aboard the *USS Santa Fe* started not by fixing the crew, but by rethinking leadership. *USS Santa Fe* was one of the worst-performing submarines in the fleet when Marquet took command in 1999. The ship suffered from poor morale, low retention, and a crew that had grown dependent on top-down orders rather than empowered decision-making. Mistakes were common. And the crew didn't feel trusted to think or act independently. Marquet recognized that these issues were systemic symptoms of a broken leadership model. To correct course, he had to shift from a culture of permission to one of ownership, giving his crew the authority and confidence to lead at every level. Marquet discovered that the most powerful change came from giving away control, developing transparency, and building competence in his Sailors. In a naval environment where transitions are constant with new technology, new missions, new leadership, the crew won't truly commit until they see a leader modeling the change. If you want better results, then start by changing how you lead. Clarity, consistency, and trust are more contagious than any directive.

Take the Helm
- **Lead the Change You Want** – When implementing change, focus first on how you lead it. Be visible, intentional, and aligned with the new direction.
- **Empower Before You Enforce** – Build ownership by giving your people authority, not just instruction.

Eyes on the Horizon
Change begins with the leader. Lead from the front and the crew will follow.

Day 212 – Change the Environment, Change the Behavior

Atomic Habits – James Clear

"Environment is the invisible hand that shapes human behavior."

James Clear explains that behavior is often a response to cues in our environment. If you want your Sailors to perform with consistency and excellence, you must first shape the surroundings that support those behaviors. Clear highlights how small shifts in environment, how tools are arranged, how routines are structured, how feedback is delivered, can lead to major shifts in behavior. This translates into division-level cultures where the workspace supports accountability, and habits are built into the rhythm of the day. That's exactly what the Navy aimed to do through the Afloat Culture Workshop (ACW) and broader Culture of Excellence initiatives. These efforts were launched to address recurring issues in command climate, toxic leadership, and performance breakdowns, not through slogans, but through intentional environmental reform. ACW sessions are working-level conversations that help commands identify toxic patterns, restructure feedback loops, and establish norms of respect, transparency, and ownership. Commands that embraced these changes saw improved retention, reduced misconduct, and greater team cohesion. Simple changes, like formalizing peer mentorship, redesigning divisional debriefs to include junior voices, or adjusting the physical setup of duty sections, had an outsized impact. When leaders change the environment with purpose, the behavior follows. Excellence doesn't just come from good intentions. It comes from better design.

Take the Helm
- **Engineer the Environment** – Evaluate the workspaces, rhythms, and tools in your division. Are they helping the new standard or holding it back?
- **Make Good Habits Easy** – Reinforce the desired behavior daily with small actions. Clarity in the environment builds consistency in execution.

Eyes on the Horizon
To change behavior, shape the deck around it. When the environment supports the standard, your Sailors will meet it.

Day 213 – Honor the Past, but Lead Forward

The Leader's Bookshelf – Admiral James Stavridis and R. Manning Ancell

"The best leaders respect tradition but are not bound by it."

Leadership requires a steady hand on the wheel and an eye on the horizon. In their book, Admiral James Stavridis and R. Manning Ancell highlight that great leaders draw wisdom from the past without becoming prisoners to it. They write, "We must learn from history but we must not be afraid to question legacy thinking when new challenges arise." Naval traditions provide a strong foundation, but leadership means knowing when to evolve the practice to meet the moment. Honoring the past means using its lessons to guide progress in an ever-evolving world. The book draws on the insights of senior leaders who emphasize balancing institutional memory with innovation. Many caution against change for its own sake but just as many warn against clinging to the familiar when the environment has shifted. Stavridis notes that "intellectual curiosity and moral courage are the traits that allow leaders to adapt while preserving what matters most." In your command, this means listening before changing, learning before leading, and then boldly moving forward with respect and purpose. History is a guide, not a leash.

Take the Helm
- **Learn Before You Lead** – When initiating change, listen first. Respect the legacy and learn why things were built that way.

- **Preserve Values, Adapt Methods** – Lead the update by connecting old strengths to new outcomes. Show your team they're building on something, not erasing it.

Eyes on the Horizon
Tradition earns respect, but the mission demands evolution. Honor what was, then lead toward what's needed now.

Day 214 – Cast the Vision Before You Chart the Course

The Infinite Game – Simon Sinek

"A just cause is a specific vision of a future state that does not yet exist. A future state so appealing that people are willing to make sacrifices to help advance toward that vision."

Great leaders ignite belief. Simon Sinek emphasizes that people are most inspired when they can see a compelling future worth striving for. He illustrates this with Dr. Martin Luther King Jr., who didn't stand before the Lincoln Memorial and outline a 10-point plan. He began with "I have a dream," a clear, emotionally resonant vision of a better future. Sinek emphasizes that King's power came from offering a just cause, a morally grounded, inclusive vision so compelling that people were willing to sacrifice, march, and endure hardship to advance it. This vision created unity before any action plan was introduced. Sinek's point is clear: belief comes before behavior. When people believe in the destination, they'll endure the journey. This principle is vital when introducing change in the Navy, whether it's a new watch rotation, a major reorganization, or a modernization effort. If you lead only with the process, then you'll only get compliance. But if you lead with the purpose, how this change contributes to warfighting effectiveness, protects Sailors, or secures long-term readiness then you'll build conviction as well. Before you chart the course, cast the vision. Let your team see what they're building, not just what they're doing.

Take the Helm
- **Connect Tasks to Meaning** – Before launching a change, explain the vision. Tie it to readiness, safety, growth, or mission impact.
- **Start With Purpose** – Communicate purpose early and often. Your clarity becomes your crew's confidence.

Eyes on the Horizon
Change without purpose breeds resistance. Cast the vision first and the course will follow.

Day 215 – Prepare for the Dip

The War of Art – Steven Pressfield

"Resistance will always rise when you're doing something that matters."

All meaningful change provokes resistance. Steven Pressfield defines resistance as the invisible force that appears any time you pursue growth or transformation. In the Navy, resistance shows up during week two of a new policy, after the first setback, or when a few voices push back louder than the rest. The mistake many leaders make is thinking something's wrong when the dip comes. In reality, the dip confirms you're on the right path. The dip creates friction, not defeat. Pressing through it leaves you stronger and changed on the other side. In the mid-19th century when the first designs for ironclads emerged, many naval officers and shipbuilders dismissed them as impractical and slow. Decades of tradition and investment in wooden sailing ships made the idea of iron-hulled, steam-powered vessels seem reckless, and support for John Ericsson's radical design, *USS Monitor*, was met with heavy resistance from within. But leaders pressed forward. They tolerated setbacks, the cultural pushback, and the critics who saw these ships as a threat to naval identity rather than its future. When *Monitor* faced off against the Confederate ironclad *Virginia* at Hampton Roads in 1862, it became clear that naval warfare had changed forever. What had seemed like a gamble was now the new standard and it all happened because committed leaders pushed through the dip.

Take the Helm
- **Press Through with Purpose** – Anticipate when energy will dip and plan extra engagement during that window.

- **Expect the Pushback** – Remind your team why the change matters. Even when progress feels slow.

Eyes on the Horizon
Resistance is part of the process, not a reason to quit. Lead through the dip, and the change will last.

Day 216 – Normalize Feedback During Change

Radical Candor – Kim Scott

"It's not cruel to challenge people. It's kind."

To sustain change, leaders must pair direction with consistent feedback. Kim Scott reminds leaders that honest, respectful feedback is a sign of care, not criticism. "Challenging directly while showing you care personally is the essence of Radical Candor," she writes. While adjusting expectations is vital in times of change, so is telling your Sailors how they measure up. Silence leaves them guessing, but feedback gives them a path to grow. Scott emphasizes that feedback should be timely, specific, and built into the rhythm of leadership. "Praise in public, criticize in private but do both consistently," she advises. This approach is central to how Sailor 360 leadership development workshops function in high-performing commands. Peer-to-peer feedback, mentorship sessions, and small-group discussions provide structured opportunities for Sailors to receive coaching in real time without stigma or formality. As Sailor 360 becomes a normal part of the command battle rhythm, so does the expectation that feedback is a tool for growth, not correction. This means normalizing feedback as a leadership habit, not a special event. When Sailors know they'll receive regular, clear input, without humiliation or delay, they engage more confidently and adjust more quickly. Change thrives where coaching is common, and accountability is consistent.

Take the Helm
- **Coach in Real Time** – Build routine, honest feedback into your change process, during and after execution.
- **Balance Candor with Care** – Model how to give and receive feedback with respect. This sets the tone for the team.

Eyes on the Horizon
Feedback is fuel. Normalize it and your team will grow into the change.

Day 217 – Communicate the Change Until You're Tired of Saying It

The Toyota Way – Jeffrey K. Liker

"Without constant reinforcement, change will fade back into habit."

Jeffrey K. Liker emphasizes that sustaining improvement requires deliberate reinforcement through communication and leadership example. He writes, "A great idea without follow-through is just wishful thinking." Don't assume that once something is said that it's understood. People need to hear it consistently, in different contexts, and from multiple voices before it becomes culture. Liker highlights Toyota's disciplined use of repetition and standardization to hardwire the behavior of entire teams. He notes, "Leaders must act as the chief change agents modeling the behavior, supporting the message, and constantly reinforcing expectations." The Navy's fleet-wide adoption of Operational Risk Management (ORM) illustrates this exact principle. ORM, first introduced in the late 1990s, aimed to change the mindset as much as the process and like any cultural shift, it didn't take hold overnight. Safety posters weren't enough. Leaders had to talk about ORM daily at quarters, before evolutions, and after drills. It became a standing item in briefs, debriefs, and leadership training. When casualties occurred, they were framed through an ORM lens: what hazards weren't identified? What controls were missed? The repetition was intentional, even exhausting, but it was the only way to embed the process into everyday thinking. If you want your message to lead change, it has to echo until it becomes second nature even when you're tired of saying it.

Take the Helm
- **Live the Message Daily** – Build your communication plan into every platform: quarters, briefs, emails, and informal walk-throughs.

- **Repeat with Purpose** – Reinforce consistently not just at rollout. Repeat the message until it becomes reality.

Eyes on the Horizon
Change sticks when it's heard, seen, and repeated. Say it often. Show it daily. Lead it always.

Day 218 – When in Doubt, Return to the Mission

The Mission, the Men, and Me – Pete Blaber

"The mission always drives the plan."

When change creates confusion or friction, clarity comes from returning to purpose. Pete Blaber emphasizes that every decision, adjustment, or action must serve the mission. That means every shift in policy, schedule, or structure should connect directly to readiness, warfighting capability, or crew safety. If it doesn't, it's noise. If it does, it's your job to lead with conviction. This mindset was foundational during the Surface Force's post-2017 training and certification overhaul. In response to catastrophic collisions and gaps in readiness, Navy leadership restructured the Surface Force Training and Readiness Manual (SFTRM) to move away from compliance-based certification and toward mission-focused proficiency. Legacy models had emphasized administrative box-checking, conducting drills or submitting paperwork, often without connection to real-world operational tasks. The new framework placed warfighting at the center of every plan. Drills were reframed not just as requirements, but as realistic scenarios tied to the ship's assigned mission areas. Success was no longer defined by completing events on time, but by demonstrating operational competency under stress. Commands that internalized this shift trained harder and with purpose. Leaders helped their crews understand that every evolution was about preparing for the next fight, not just passing an inspection. Blaber's central lesson held true: when leaders anchor change to the mission, Sailors respond with urgency, trust, and buy-in.

Take the Helm

- **Refocus During Friction** – In conflict or confusion, ask: How does this serve the mission? Use the answer to guide your decisions.

- **Anchor Every Change in Purpose** – Help your Sailors see their role in the bigger picture. Purpose reduces resistance.

Eyes on the Horizon

The mission cuts through the noise. Lead your team back to purpose, and they'll find their direction.

Day 219 – Let Your Team Lead the Change

The Dichotomy of Leadership – Jocko Willink and Leif Babin

"When you empower others to lead, they take ownership and that's when change sticks."

Lasting change happens when the team takes the wheel. This means creating space for Sailors to solve problems and carry the vision forward without constant direction. If the change only moves when you push, then that is dependence, not leadership. That principle holds true not only on the deckplates, but at the very roots of naval history. In 1798, the creation of the Department of the Navy was not the product of a sudden top-down reform, but the culmination of years of advocacy from operational leaders, ship captains, commercial seafarers, and political allies who saw firsthand the dysfunction of managing naval affairs under the War Department. They weren't passive recipients of policy. They were the voices of the mission. Congress eventually responded, not because a new department was convenient, but because the people closest to the fleet made the strongest case for change. Once established, the Department of the Navy empowered naval leadership to shape doctrine, direct resources, and define its own professional standards. It institutionalized ownership because the people who would carry out the change were the same people who helped drive it. That's the lesson that endures: real change takes hold not when it's handed down, but when it rises from within.

Take the Helm
- **Step Back to Build Up** – Identify natural leaders and let them take the reins. Support them visibly and give them room to fail and grow.
- **Create Ownership Opportunities** – Celebrate progress that comes from the deckplate not just the plan.

Eyes on the Horizon
Change driven by leadership is effort. Change driven by the team is momentum. Empower both and watch it grow.

Day 220 – Know When to Pivot

The Lean Startup – Eric Ries

"The ability to pivot is the hallmark of a resilient organization."

In fast-paced environments, rigidity breaks, and adaptability survives. Eric Ries explains that successful organizations do not cling blindly to their first plan. Each initiative becomes an experiment to learn and improve. One that is measured, tested, and adjusted as new feedback comes in. In the Navy, new programs, changes in watch rotations, or process overhauls all require flexibility. That mindset was reflected in the 2018 reactivation of U.S. 2nd Fleet. Originally deactivated in 2011 as part of post-Cold War restructuring, 2nd Fleet had long been responsible for command and control in the North Atlantic. At the time, consolidating its functions under Fleet Forces Command seemed efficient, but within a few years, global dynamics shifted. Russian submarine activity increased, gray-zone maritime threats rose in the Atlantic, and cyber disruptions emerged as a strategic concern. Rather than cling to an outdated model, the Navy pivoted by reviving 2nd Fleet to provide focused operational leadership for the Atlantic theater. It was a forward-looking decision that acknowledged new conditions, restructured command roles, and restored fleet-level oversight where it was operationally needed. The core mission, maritime security and forward presence, didn't change. But how the Navy organized to meet that mission did. When the environment evolves, so must your method. Resilience comes from responsiveness.

Take the Helm
- **Listen for the Pivot Point** – Pay attention to friction, feedback, and performance. Adapt early, not after failure.

- **Change the Approach, Not the Intent** – Be willing to adjust without losing the vision. Flexibility earns credibility.

Eyes on the Horizon
Being able to adapt demonstrates wisdom far more than rigid stubbornness does. Pivot when needed, and the change will remain strong.

Day 221 – Lead Stability Through Transition

Call Sign Chaos – Jim Mattis and Bing West

"In a time of transition, a leader must be a shock absorber."

Whether stepping into command, assuming new responsibilities, or navigating a shift in mission, Jim Mattis and Bing West underscores the value of steady presence during transition. Sailors instinctively look to their leaders during change, which means that your tone, presence, and consistency become the anchor that steadies the ship while the environment adjusts. Mattis writes, "A leader must be calm and clear-headed in the face of uncertainty" and describes how composure under pressure is about projecting control and clarity. He emphasizes that in the most chaotic situations, the leader's job is to impose order, not mirror the noise. Following the attacks of Sept. 11, 2001, as the Navy rapidly shifted from peacetime operations to a global warfighting posture, leaders at all levels had to guide their Sailors through uncertainty, extended deployments, and an evolving threat environment. The leaders who steadied their watch teams, reinforced the mission, and communicated clearly through each change served as shock absorbers for their crews. Whether during a change of command or a shift in triad leadership, the way you carry yourself becomes a signal to your team. When you stay grounded, they find their footing too. That's not just managing change, but rather it's leading through it.

Take the Helm
- **Project Calm During Transition** – During transitions, double down on presence, predictability, and tone. Your steadiness shapes theirs.
- **Communicate With Consistency** – Acknowledge change but reinforce continuity. Remind your team what won't change.

Eyes on the Horizon
Be the constant when everything else shifts. That's what earns trust and keeps the team moving forward.

Day 222 – Small Wins Make Big Change Possible

Mindset – Carol S. Dweck

"Effort is what ignites ability and turns it into accomplishment."

Carol S. Dweck explains that people with a growth mindset see progress as the product of consistent effort, not innate talent or sudden transformation. She writes, "It's about learning, improving, and pushing forward, one step at a time." This means building change not through grand gestures, but through repeatable, visible wins that strengthen belief and momentum. Dweck emphasizes that celebrating effort and progress is what turns goals into habits. "The path to success," she notes, "is paved with small goals, consistently met." Leaders who recognize and reinforce those small wins build not only competence, but confidence. That belief fuels future effort and eventually, transformation. The Navy's phased integration of female Sailors into the submarine force reflects this mindset in action. Rather than attempting a fleetwide overhaul, the Navy began with a small, deliberate step, assigning a handful of female officers to select ballistic missile submarines with full command support and careful logistical planning. By focusing on professionalism, inclusion, and operational excellence in these early assignments, leaders created a foundation of trust and credibility. Those initial successes built momentum, opened doors for enlisted women, and gradually reshaped what was once thought impossible. If you want your Sailors to believe in a change, show them what success looks like early and often. One win becomes two. And two become a culture shift.

Take the Helm
- **Build with Small, Visible Wins** – Design small, achievable changes that build momentum. Track them publicly and celebrate consistently.

- **Celebrate the Effort, Not Just the Outcome** – Reinforce that steady effort leads to improvement even if it's not immediate.

Eyes on the Horizon
Big change starts small. Win today, and your team will believe in what's possible tomorrow.

Day 223 – Lead With Optimism, Not Spin

The Infinite Game – Simon Sinek

"Optimism is not the denial of reality. It is the belief that the future is bright even if the present is difficult."

When leading through change, your Sailors need honesty paired with hope, not hollow motivation. The kind of optimism that doesn't ignore the hardship but frames it with purpose. Simon Sinek reminds us that real optimism acknowledges difficulty while affirming the team's ability to overcome it. This kind of transparency builds trust, especially when the situation is uncertain or morale is strained. The strongest leaders speak plainly about challenges, expectations, and progress, because they know clarity, not spin, is what fosters resilience. This mindset is especially critical during extended maintenance availabilities and shipyard periods, when the mission feels distant and routines are disrupted. Sailors feel the weight of changing schedules, operational delays, and the personal toll on their families. In these seasons, your presence and your words matter most. Leaders who acknowledge the stress while reinforcing the long-term readiness being built help Sailors see beyond the daily grind. They tie every compartment inspection, every rehab plan, every watchstanding detail back to warfighting capability. That connection turns frustration into ownership. When missions shift, workload spikes, or energy dips, it's your tone that shapes the climate. Don't sugarcoat reality. Don't spin the story. Own it, frame it, and give your team a reason to stay in the fight. That's optimism that earns loyalty.

Take the Helm
- **Be Honest and Hopeful** – Speak truthfully about the challenges of change. Then reinforce what makes it worth doing.
- **Lead With Clarity, Not Comfort** – Speak directly. Your team will respect your honesty more than empty reassurance.

Eyes on the Horizon
Honest optimism inspires action. Lead with truth and belief and your Sailors will rise to meet the moment.

Day 224 – Stay Consistent When Others Get Uncomfortable

Discipline Equals Freedom – Jocko Willink

"You can't let your feelings dictate your actions. Discipline drives the standard."

Leadership during change demands unshakable discipline. Jocko Willink reminds us that when resistance builds, consistency becomes the weapon that carries change forward. That means holding the line when new routines feel uncomfortable or unpopular, because your commitment becomes the compass for everyone else. Leaders who waver under pressure send the signal that standards are negotiable, but those who remain calm and focused build cultures where accountability spreads. Nowhere was this more visible than during the Battle off Samar in 1944, when the vastly outgunned U.S. task unit Taffy 3 faced a sudden attack by the full weight of the Japanese Center Force. Commander Ernest Evans of *USS Johnston* had no illusions about the odds but he never paused to poll the room or soften the blow. He pushed forward without hesitation, upholding the standard of combat readiness and courage his crew had trained for. Even when others hesitated, Evans led from the front, reinforcing discipline not with punishment, but with personal example. His refusal to yield gave his crew permission to fight with everything they had, and their collective discipline firing torpedoes, laying smoke, and pressing the assault, shattered enemy momentum. Change is uncomfortable. Crisis is worse. But Sailors don't need comfort. They need consistency.

Take the Helm

- **Be Relentlessly Consistent** – Stick to the plan when resistance rises. Reaffirm the standard with discipline, not emotion.

- **Lead Through Repetition** – Use consistency as a leadership message: "This matters. I'm not backing off."

Eyes on the Horizon
When others waiver, your discipline is the anchor. Stay steady and the team will follow.

Day 225 – Align the Mess, the Wardroom, and the Deckplates

The Five Dysfunctions of a Team – Patrick Lencioni

"If you want the team to function, the leadership must be united."

No change effort succeeds without alignment at every level of leadership. Patrick Lencioni explains that a lack of trust or inconsistent messaging at the top fractures the team below. That means the Chief's Mess, the Wardroom, and the deckplate leaders must present a united front of purpose and tone. This principle was tested across two of the Navy's most significant public health crises: the 1918 influenza and the 2020 COVID-19 pandemics. During the influenza outbreak, the Navy was still expanding rapidly due to World War I. Despite limited medical understanding, most commands enforced unified sanitation and isolation procedures, often backed strongly by both officers and Chiefs. As a result, many Navy installations saw lower mortality rates than surrounding civilian areas. By contrast, during the COVID-19 pandemic, despite modern medicine and communication systems, some commands struggled to maintain consistency. Misalignment between senior leadership and medical staffs on some ships led to conflicting guidance, crew confusion, and fractured trust. Yet other ships that deliberately aligned the Mess, Wardroom, and LPOs in tone and action saw smoother implementation of quarantine and operational pivots. The difference was solidarity. When leaders speak with one voice, reinforce the same values, and model the behavior they ask for, the crew follows, not just with compliance, but with conviction.

Take the Helm
- **Establish a Shared Front** – Before launching change, build alignment across leadership layers then speak with one voice.
- **Close the Gaps in Leadership** – Address disagreements behind closed doors, and present unity in front of the team.

Eyes on the Horizon
If your leadership team isn't aligned, your Sailors won't be either. Build unity first then lead the change together.

Day 226 – Change Moves at the Speed of Trust

Leaders Eat Last – Simon Sinek

"Trust is built on telling the truth, not telling people what they want to hear."

Lasting change starts with trust. Simon Sinek emphasizes that trust is built in the everyday actions of leaders who show up, listen, and lead with integrity. When Sailors see that your decisions are grounded in care, clarity, and shared purpose, they lean into the change even when it's hard. But without trust, even the best ideas will stall. That truth has been embedded in the Navy's DNA since April 1, 1893, when the rank of Chief Petty Officer was officially established. But the promotion wasn't just about adding structure, it was about formally recognizing trust that had already been earned on deckplates for generations. Their authority came not from collar devices, but from years of guiding junior Sailors, advising officers, and maintaining the heartbeat of the ship. That trust became the foundation of an enduring leadership institution. Even today, a new process, standard, or cultural shift aboard ship often passes through one critical filter: "Do the Chiefs believe in this?" Because when they do, when they trust the change and the team trusts them, everything moves faster, stronger, and stands ready under pressure.

Take the Helm

- **Invest Before You Implement** – Build credibility before enforcing change. Start with relationships, follow through with action.

- **Lead with Consistency and Transparency** – If trust is low, move slower and communicate more. Your people need to know your intentions before they follow your direction.

Eyes on the Horizon
Change moves with trust, not by force. Earn it first, and the rest will follow.

Day 227 – Change Requires Follow-Up, Not Just a Kickoff

Good to Great – Jim Collins

"Greatness is not a function of circumstance. Greatness, it turns out, is largely a matter of conscious choice and discipline."

Big change rarely fails at the starting line. Most breakdowns happen after the kickoff when momentum fades and leaders stop reinforcing the standard. Collins emphasizes the power of the "flywheel effect," small, repeated actions that build momentum until breakthrough. If you want a new standard to take root, you must revisit it, reinforce it, and reflect it in your own leadership. The Littoral Combat Ship (LCS) program is a study in the importance of this principle. Designed to revolutionize close-to-shore operations, LCS promised speed, modularity, and reduced manning through automation and mission-flexible configurations. The kickoff was bold, but the initial vision wasn't matched by sustained execution. The Navy struggled with rotating crew models, integrating modular mission packages, maintaining readiness, and defining the ship's operational purpose. But rather than abandon the platform, the Navy responded with course corrections, repurposing LCS for evolving missions like mine countermeasures and maritime security. The program's value is being realized in the resilience of the Sailors who've worked to align its vision with operational reality. Despite early setbacks, the LCS program has demonstrated operational value through successful deployments like *USS Gabrielle Giffords* with her successful launch of a Naval Strike Missile and effective counter-narcotics missions across multiple fleets. The flywheel may have turned slower than expected, but it continues to turn. Proof that disciplined follow-through can still shape lasting capability.

Take the Helm
- **Integrate, Don't Abandon** – Build checkpoints into your leadership rhythm to follow up on change progress.

- **Reinforce the Rollout** – Hold your leaders accountable for reinforcement. Momentum comes from repeated presence.

Eyes on the Horizon
Change takes root through persistent follow-up, long after the kickoff is over. Stay involved, and your team will stay engaged.

Day 228 – Let Go of What No Longer Serves the Mission

The Art of War – Sun Tzu

"When the situation changes, adjust your strategy."

Sun Tzu emphasizes that the key to strategic success is flexibility in the face of changing circumstances. "Do not repeat the tactics thatve gained you one victory," he writes, "but let your methods be regulated by the infinite variety of circumstances." It's tempting to keep doing things "the way they've always been done." But when the mission evolves, leaders must have the courage to let go of routines, systems, or mindsets that no longer deliver. For decades, battleships were the embodiment of naval power, floating fortresses with massive guns, armor, and prestige. But as air power evolved and naval aviation matured, the strategic landscape changed. Early carrier advocates, like Admiral William Moffett, faced resistance from traditionalists who saw carriers as auxiliary platforms rather than capital assets. It wasn't until the Pacific War, and especially the stunning success of carrier-led operations at Midway and Coral Sea, that the Navy fully embraced the shift. Battleships, once central to naval doctrine, were relegated to support roles or left behind entirely. Carrier strike groups became the new cornerstone of American sea power. That transition didn't come easily but it came because leaders recognized that clinging to the past would cost the future. Abandoning the battleship design marked a shift from tactics to clear, decisive strategic foresight.

Take the Helm
- **Audit the Routine** – Identify policies, habits, or routines that no longer support the mission then initiate discussion about updating or retiring them.

- **Adapt Without Attachment** – Reinforce that letting go isn't quitting. Letting go shows the courage to evolve and embrace what's next.

Eyes on the Horizon
Real change requires embracing new ideas and clearing away what no longer fits. Make room for what the mission truly requires.

Day 229 – Change What You Measure, Change What You Value

The Toyota Way – Jeffrey K. Liker

"You get what you inspect, not what you expect."

The values of a team are revealed by what gets measured. This means leaders must be intentional about how they evaluate progress. Why? Because what gets measured becomes a signal of importance. When leaders reinforce that critical thinking, resilience, and team cohesion count just as much as raw output, Sailors adapt to meet those expectations. People respond to what is tracked because it defines success in the eyes of the institution. Over time, the scoreboard becomes the culture, elevating performance and identity. Jeffrey K. Liker shows how Toyota's strength lies in measuring the right things, favoring long-term value over short-term output, depth over speed, and team performance over individual checkmarks. This principle is evident in the Navy's integration of Warrior Toughness into boot camp and officer accession pipelines. Instead of measuring success solely through physical fitness and academic scores, these programs introduced new metrics around spiritual resilience, mental toughness, and alignment with Navy values. The message resonated: real toughness means building up the whole Sailor, not just their physical strength. By elevating these dimensions through deliberate measurement, the Navy began to reshape how it defines readiness and leadership potential. If you want cultural change, start by asking, "Are we measuring what really matters?" Because once you shift the metrics, you shift the mindset and that's when transformation begins.

Take the Helm
- **Reevaluate the Scoreboard** – Review your current measures of success. Do they reflect what the team truly needs to value now?

- **Measure for Meaning, Not Just Motion** – Make metrics visible and tied to change and progress, not just performance.

Eyes on the Horizon
Change the scoreboard, and you'll change the game. Lead your team to measure what really matters.

Day 230 – Reinforce Change Through Recognition

How to Win Friends and Influence People – Dale Carnegie

"Be hearty in your approbation and lavish in your praise."

Change sticks when people feel seen. Dale Carnegie teaches that recognition is a force multiplier. "People will do more for a sincere compliment than for money," he writes. When leaders acknowledge effort, progress, and attitude, they reinforce exactly the behaviors that build momentum. Carnegie emphasizes that praise must be specific, timely, and authentic. "Give honest and sincere appreciation," he advises. General flattery falls flat but thoughtful recognition motivates action. The Navy's Battle Effectiveness Award program reflects this principle in action. Commands that embrace high standards across warfare areas and consistently demonstrate readiness, innovation, and sustained excellence are formally recognized with the Battle "E." Such recognition means the command's drive for excellence and improvement is visible, appreciated, and celebrated, not just displayed in paint or metal. For crews navigating change, this kind of recognition reinforces that the hard work matters. When Sailors adapt early, go the extra mile, or reinforce the new standard for others, call it out. Not as a formality, but as a message: this is what right looks like.

Take the Helm
- **Make Praise Personal and Public** – Look for early adopters and quiet influencers. Recognize their role in reinforcing the change.

- **Celebrate Visible Effort** – Build recognition into your weekly rhythm. What you celebrate shapes what sticks.

Eyes on the Horizon
Praise is a leadership multiplier. Use it to fuel the change and spotlight the culture you want to build.

Day 231 – Make It the New Normal

Extreme Ownership – Jocko Willink and Leif Babin

"There are no shortcuts. There's only the discipline to do it right and keep doing it right."

Change becomes culture the moment it stops feeling like change. After the briefs are delivered and the rollout ends, the true challenge begins: embedding the new expectation so deeply that it becomes second nature, no longer requiring justification or reminder. Discipline is what makes new procedures feel familiar. That principle was put to the test after the devastating fire aboard *USS Forrestal* in 1967, which claimed the lives of 134 Sailors and exposed severe gaps in shipboard damage control readiness. The Navy responded by overhauling its approach and instilling the doctrine of "Every Sailor a Firefighter." What began as a corrective action became a cultural anchor and second nature across the fleet. Sailors don't need to be reminded why it matters, because it's already part of who they are. Leaders stopped treating firefighting as a special initiative and started modeling it as the baseline standard. When leaders stop treating the change as a temporary fix and start modeling it as the new default, the crew follows. You don't need to declare that a new process has arrived, you just need to live it like it always belonged. Consistency turns effort into habit, and habit into identity. That's when change takes root, not just in policy, but in culture.

Take the Helm
- **Treat the Change as the Standard** – Stop calling it "new" after the rollout. Use language that reinforces it as the current and correct way.
- **Reinforce Through Routine** – Model consistency daily. What's routine for you becomes routine for your team.

Eyes on the Horizon
Change becomes culture through repetition. Keep leading it until no one remembers it was different.

Day 232 – Don't Expect Buy-In Without Buy-In

Crucial Conversations – Kerry Patterson, Joseph Grenny, Ron McMillan, Al Switzler

"People don't get defensive because you disagree. They get defensive because they don't feel heard."

Change imposed is change resisted. Leaders often underestimate how critical it is for people to feel heard, especially during transition. In the Navy, where directives can come fast, rank can unintentionally silence honest feedback and if Sailors feel ignored or sidelined, even the most well-reasoned plans will meet resistance. The repeal of "Don't Ask, Don't Tell" in 2010 is one of the clearest examples of this principle at scale. Before allowing LGBTQ+ service members to serve openly, the Department of Defense launched the largest internal listening campaign in military history, surveying over 115,000 active-duty members and 44,000 military spouses. Across fleets and bases, Sailors and Marines were asked how the change would affect unit cohesion, readiness, and morale. Town halls were held, commanders were trained, concerns were acknowledged without dismissal, and implementation timelines were shaped around what was learned. The result was one of the smoothest social transitions in the modern military. Commands that embraced the conversation early saw trust deepen and those that didn't had to work harder to catch up. The message to every leader was clear: if you want people to get on board, invite them to the table before you start sailing.

Take the Helm
- **Invite the Conversation Early** – Before enforcing a change, ask your team how it will affect them. Listen fully before responding.
- **Listen Like It Matters** – Use small adjustments to show your team they've been heard. Demonstrated respect is a powerful motivator.

Eyes on the Horizon
People resist change they don't feel part of. If you want buy-in, then invite ownership from the start.

Day 233 – Keep Change Simple Even When It's Hard

The 7 Habits of Highly Effective People – Stephen R. Covey

"The main thing is to keep the main thing the main thing."

Complexity is the enemy of execution. Stephen R. Covey reminds leaders that clarity and focus are the foundation of lasting success. He writes, "When priorities are clear, decisions become easier." During change, your Sailors may face shifting routines, new technologies, or uncertain expectations. Your role is to cut through the noise, identify what matters most and keep that message consistent. Covey teaches that highly effective leaders operate with a principle-centered mindset. They avoid overloading their teams with excessive detail or scattershot priorities. Instead, they define the overarching priority that everyone's focus should orbit, then align daily actions to that purpose, and reinforce it with discipline. During Operation Tomodachi in 2011, following the earthquake and tsunami in Japan, the Navy's humanitarian response required coordination across multiple commands in a rapidly changing environment. Leaders maintained focus by consistently reinforcing one clear priority: provide aid to the Japanese people and ensure the safety of the crew. That singular message cut through the operational complexity and kept Sailors grounded in their mission. Covey notes, "Effective leadership is putting first things first." Simplicity can be a tactical advantage. When your team is overwhelmed, simplify the objective, not the standard. That's how you keep them moving forward.

Take the Helm
- **Simplify Execution Without Lowering the Bar** – Filter every change through one question: "What's the simplest way to execute this without losing quality?"
- **Clarify the Mission Daily** – Reinforce the core purpose and essential tasks. Eliminate distractions that dilute the focus.

Eyes on the Horizon
Clarity beats complexity every time. Keep it simple, and your team will stay in the fight.

Day 234 – Change Is a Constant; Lead Like It

Sea Power – Admiral James Stavridis

"The sea is never still. And neither is the world."

In maritime life, change is the environment. Admiral James Stavridis reflects on centuries of naval history to illustrate one truth: the ocean is always shifting, and effective leaders adapt with it. He emphasizes that in today's Navy, change comes in many forms: technology, geopolitics, missions, and personnel. The best leaders trade frustration for readiness when challenges arise. From the rise of the aircraft carrier to the post-9/11 reorientation toward asymmetric threats, *Sea Power* traces how the U.S. Navy's greatest leaps forward came not from resisting change but from steering into it. Stavridis recounts how forward presence in the Mediterranean evolved during the Cold War, how naval forces adapted to humanitarian crises like the 2004 Indian Ocean tsunami off the west coast of Sumatra, Indonesia and the 2010 Haiti earthquake, and how emerging cyber threats reshaped maritime security. He highlights that leaders who thrived in these transitions did so not by clinging to old tactics, but by staying curious, responsive, and strategically flexible. He notes, "The greatest captains don't wait for the storm to pass. They adjust the sails." That mindset must shape your leadership. Don't treat change as a one-time challenge. Build your team to expect it, embrace it, and stay steady through it. When change becomes normal, progress becomes possible.

Take the Helm
- **Lead with a Steady Hand** – Set the expectation with your team that change is part of the mission, not a disruption to it.
- **Equip Your Team to Adapt** – Coach adaptability the same way you coach readiness. It's a core competency.

Eyes on the Horizon
Change charts the course and keeps the mission moving forward. Lead like it's normal, and your team will navigate with confidence.

Day 235 – Lead with Presence, Not Just Policy

Leadership Strategy and Tactics – Jocko Willink

"The most important part of leadership is not what you say, it's what you do."

Showing up for your team builds more leadership than any memo ever could. Jocko Willink reminds us that your team doesn't just listen to what you say, they watch what you model. He writes, "If you want them to care, you have to care first. If you want them to engage, you have to be engaged." During change, even the best policy will falter if the leader disappears. But when you're visible, involved, and invested, your Sailors gain the confidence to push through uncertainty. Willink emphasizes that the strongest leaders are on the ground, not hovering above it. "When your people see you with them, they know they're not alone," he says. That presence matters most during high-stakes evolutions like pre-deployment readiness checks and final certification events. When leaders walk the deckplates during light-offs, general quarters, and final drills, not to inspect but to support, they reinforce trust and show the crew that the mission is shared. More than oversight, these moments cultivate ownership and remind Sailors that leadership remains present and engaged. Showing up means more than walkthroughs or briefs, it means being present where the work is happening, listening without agenda, and reinforcing the mission with steady actions. Policy sets the plan. Presence earns the trust that makes it work.

Take the Helm

- **Lead Where It Matters** – Increase your visibility during periods of change. Be where the work is.

- **Turn Policy into Action** – Reinforce messages with presence, not just memos. Your walk builds belief.

Eyes on the Horizon

You can't lead change from the sidelines. Show up, stay involved, and your team will follow you forward.

Day 236 – Change Is Leadership in Action

Turn the Ship Around! – L. David Marquet

"Leadership is not about giving orders. It's about giving control."

Real leadership drives change forward instead of standing still. L. David Marquet shows that the most powerful transformation happens when leaders stop giving commands and start giving control. He writes, "When you give people control, you create leaders." His own experience commanding the *USS Santa Fe* proved this firsthand. By shifting from a top-down command style to an intent-based leadership model, Marquet empowered his crew to think, decide, and lead. The result was a cultural overhaul. *USS Santa Fe* went from the bottom of the fleet to one of the best-performing submarines in the Navy, not because of stricter orders, but because the crew owned the change themselves. Leaders who leave a lasting mark are the ones who empower their people to adapt, decide, and grow. Marquet teaches that sustainable change comes not from top-down direction, but from bottom-up ownership. Trust your Sailors to lead and think for themselves, and they'll become the ones who carry change through the fleet. That's the difference between a crew that complies and one that commits. Your ability to guide that process by building trust, reinforcing discipline, and releasing control is what defines your leadership. Change isn't what gets in the way of leadership. It *is* leadership when done with intention and purpose.

Take the Helm
- **Empower Ownership** – Empower your Sailors to own the future, not just the checklist. Teach them to lead the next change.

- **Lead Through the Shift** – Reflect on how you led not just what you achieved. Your process becomes your signature.

Eyes on the Horizon
Change is the proving ground of leadership. Lead it with integrity and your impact will echo far beyond your tour.

Day 237 – Change Demands Patience and Pressure

Great by Choice – Jim Collins and Morten T. Hansen

"The best leaders exert consistent pressure, not erratic intensity."

Greatness isn't built in sprints, but rather forged in disciplined, sustained motion. Jim Collins and Morten T. Hansen introduce the principle of the "20 Mile March," a metaphor for consistent effort even when conditions vary. Change often starts fast but if it isn't reinforced steadily, it fades just as quickly. Real leadership means resisting the urge to chase quick wins and instead applying patient, purposeful pressure over time. That pressure matters because change is rarely welcomed without friction. You'll face cultural inertia, skeptical voices, competing priorities, and the natural pull back toward comfort and familiarity. Pressure then is about presence. It's the leader continuing to ask the same questions, enforce the same standard, and reinforce the same vision even when enthusiasm dips or early progress stalls. It's what keeps momentum alive when the excitement of the rollout wears off. Collins and Hansen show that without steady leadership pressure, organizations don't just slow, they regress. For Sailors, that means leadership must show up daily not just with demands, but with presence, follow-up, and accountability. New standards take hold when leaders keep showing up, reinforcing them until they become routine. Progress, when measured over time, proves more powerful than any one push.

Take the Helm
- **Lead with Relentless Consistency** – Set consistent expectations and timelines. Don't rush, but don't let up.
- **Pace the Push** – Use routine engagement to apply steady pressure. Track progress publicly, celebrate small wins, and revisit priorities often to keep change visible and valued.

Eyes on the Horizon
Change needs time but also tension. Apply both with discipline, and the result will last.

Day 238 – Build Change That Outlasts You

Legacy – James Kerr

"True leaders plant trees they'll never sit under."

James Kerr explores how lasting impact comes from leaders who focus not on personal recognition, but on setting others up to succeed long after they're gone. "Be a good ancestor," he writes. Meaning the true test of leadership shows in how things run after you've handed over the watch. When change disappears after you leave, that shows it never became culture, only enforced compliance. This is why the commands that successfully integrate Warrior Toughness into their training pipelines see real cultural traction. Warrior Toughness, a Navy initiative that strengthens Sailors' mental, physical, and spiritual resilience, was designed to help them thrive under pressure both on and off duty. Because leaders train their teams to carry this resilience forward, Warrior Toughness doesn't fade with turnover, but endures because it is owned. Kerr notes, "Legacy is not what you leave for people, it's what you leave in people." In your command, that means building systems others can carry, mentoring leaders who can rise, and instilling standards that live beyond your signature on the plan of the day. That's how change becomes culture and leadership becomes legacy.

Take the Helm

- **Grow the Next Leaders** – Develop Sailors who can lead the change when you're gone. Mentor them early.

- **Design for Continuity** – Build structure and documentation so the change is repeatable, transferable, and sustainable.

Eyes on the Horizon
What defines a leader's legacy is less about personal deeds and more about what lasts in their absence. Build to last.

Day 239 – Change Happens One Conversation at a Time

Crucial Conversations – Kerry Patterson, Joseph Grenny, Ron McMillan, Al Switzler

"The key to real change lies in the conversations we're not having."

One honest discussion does more to shift a culture than a hundred all-hands briefs. The authors emphasize that culture shifts when leaders engage in high-stakes conversations with care, clarity, and courage. They write, "The people who are most influential are those who master crucial conversations." That means stepping beyond the podium and into the one-on-one conversations that build trust, clarify expectations, and move people forward. The book highlights that silence, avoidance, and vague direction are the silent killers of change. Real leadership shows up in how you speak when tension is high, emotions run strong, or doubt sets in. This shows up clearly in command career counselor boards, where honest retention conversations have far greater impact than blanket incentives. Leaders who sit down with Sailors to understand their goals, concerns, and motivations create space for commitment and trust to grow. "Speak persuasively, not abrasively," the authors advise, "and listen as if your life depended on it." One real conversation can retain a Sailor far more effectively than a policy ever could.

Take the Helm
- **Have the Conversations That Matter** – Identify key influencers and change agents and have direct, honest conversations to gain their alignment.

- **Engage Through Everyday Conversations** – Use casual moments like walk-throughs, check-ins, mess decks to reinforce the change with real dialogue.

Eyes on the Horizon
Change becomes real in conversation. Don't lead with announcements. Lead through connection.

Day 240 – Celebrate What the Team Became

With the Old Breed – Eugene B. Sledge

"Leadership through change is about more than milestones. It's about what the journey makes of your team."

Eugene B. Sledge, reflecting on his experiences as a Marine mortarman, some of the most brutal and defining World War II battles, reflects not just on the battles themselves, but on the bond, resilience, and transformation forged through hardship. "We had become brothers in the most profound sense of the word," he writes. The struggle was brutal, the environment unforgiving, but the character that emerged defined them for life. This same transformation was seen in the Navy Corpsmen and Seabees who deployed alongside Marines in Fallujah and Ramadi during the height of the Iraq War. Under constant danger, far from familiar roles, these Sailors forged deep, lasting bonds through grit, sacrifice, and unshakable commitment to one another. What they endured together became their identity. Change efforts aren't always heroic, but they are formative. What your Sailors overcome together matters just as much as what they achieve. Sledge's story shows that true pride grows not from inspections or awards but from what you and your team endured side by side. The growth, the grit, the shared sacrifice, those are the moments that define a team's identity. As a leader, take time to acknowledge not just what was done, but who they became through it. When people see that their effort changed something within, not just around, they own that change for life.

Take the Helm

- **Reinforce Identity Through Reflection** – After major change efforts, pause to reflect as a team. Highlight the growth, not just the gains.

- **Honor the Journey, Not Just the Outcome** – Use ceremonies, letters, or recognition to mark the journey and reinforce identity.

Eyes on the Horizon
What your Sailors became during the change is the real win. Celebrate it and they'll carry it forward.

PART III:

THE ENDURING LEADER

CHAPTER 9:
THE WARFIGHTER'S MINDSET

Day 241 – The Fight Finds You

Extreme Ownership – Jocko Willink and Leif Babin

"Combat is reflective of life, only amplified and intensified."

You don't always choose the moment, but as a leader, you must be ready when the moment chooses you. Jocko Willink and Leif Babin describes leading SEAL Team Three's Task Unit Bruiser during the intense urban combat of the Battle of Ramadi. Decisions came fast, under fire, and with lives on the line. There was no time to wish for clarity or wait for ideal conditions. Leadership meant taking immediate ownership, staying calm, and executing with confidence even amid chaos. The battlefield punished hesitation. So does leadership. Whether in combat or on the deckplates, you must cultivate the discipline to act without delay when the situation demands it. During one operation in Ramadi, Willink recounts how a friendly Iraqi soldier was mistakenly killed in a blue-on-blue incident. The moment demanded immediate control of the team, of the confusion, of the consequences. Willink didn't wait for someone else to accept responsibility. He owned the failure and used the incident to drive home the critical importance of clear communication and decisive leadership.

Take the Helm
- **Train for the Unknown** – Conduct mental rehearsals before every high-risk evolution or transit. Imagine worst-case scenarios and commit to how you'll lead.

- **Commit to Action** – In high-readiness watches or combat exercises, emphasize mental presence as much as technical knowledge. Prepare your Sailors to act without waiting.

Eyes on the Horizon
You may not choose the fight, but you must be ready when it finds you. Leadership under pressure is won in moments you pre-committed to lead long before they arrive.

Day 242 – Aggression Is a Leadership Weapon

The 33 Strategies of War – Robert Greene

"The best defense is not retreat or compromise, but overwhelming force and constant pressure."

An aggressive leader acts purposefully, claims the initiative early, and keeps fear from dictating when and how to move. It's deliberate action in the face of uncertainty. Robert Greene emphasizes that success often goes to the leader who seizes momentum, applies pressure, and forces others to react. Waiting for perfect clarity can surrender initiative, but boldness, when grounded in intent, creates tempo and confidence. In 1950, U.S. and U.N. forces faced a deteriorating situation in Korea. Rather than settle into defensive posturing, the Navy helped execute General MacArthur's audacious plan: an amphibious assault at Incheon, far behind enemy lines. Naval forces aggressively controlled the waterways, cleared mines, and landed Marines in a narrow tidal window with precision and speed. The boldness of the operation caught North Korean forces completely off guard and turned the tide of the war. Aggression, when paired with preparation and purpose, is a leadership weapon. Whether recovering from a casualty, leading a boarding team, or taking initiative in uncertain conditions, your ability to act with speed and resolve defines the pace of your team. Passive leadership waits and lets problems grow. Aggressive leadership attacks problems head-on and wins.

Take the Helm

- **Make Aggression a Skill, Not a Reaction** – Teach your team the difference between informed aggression and carelessness. Aggression is about decisive momentum, not uncontrolled action.

- **Seize the Initiative** – Use combat team trainers and real-world examples to show how initiative under pressure wins battles and builds trust.

Eyes on the Horizon

The hallmark of an aggressive leader is taking decisive action without needing to be told when to move. When action is needed, the team looks to the first voice with clarity and courage.

Day 243 – Train Like It's Real Because It Might Be

The Unforgiving Minute – Craig Mullaney

"In combat, there's no such thing as a perfect plan. Only preparation, discipline, and the ability to adapt faster than your enemy."

Combat reveals every gap in preparation. Craig Mullaney recounts how, despite excelling in academic and theoretical environments, his real education came during close quarters combat in Afghanistan when stress and violence stripped away ego and exposed only what had truly been rehearsed. He and his team tested this principle when they encountered enemy fighters in terrain that nullified their planned tactics. His unit was climbing a ridge in the rugged mountains near Asadabad when they were pinned down by coordinated enemy fire. Radio comms were shaky and visibility was almost nonexistent. It was Mullaney's first firefight, a moment no amount of classroom preparation could simulate. In that instant, there was no time to strategize or deliberate. The team's survival depended on immediate action drills, fire discipline, and the battle rehearsals they had internalized back at Fort Drum. Ultimately, the platoon was able to withdraw with support from air and artillery assets, but the emotional and operational toll of that day left a permanent mark. Mullaney carried this truth into every mission that followed: in the absence of experience, you will fall back on your training and that training must be real.

Take the Helm
- **Demand Realism** – During every drill, ask yourself: "If this were real, would we survive?" Then adjust your training accordingly.
- **Set the Tone Early** – Your crew will mirror your seriousness during training. If you treat it like it matters, so will they.

Eyes on the Horizon
There is no more practice when the fight begins. Train your team for what's coming, not just what's comfortable.

Day 244 – Calm Is a Combat Multiplier

The Mission, the Men, and Me – Pete Blaber

"The battlefield is loud. Leaders must be the quietest, clearest voice in the chaos."

More than a trait, calm acts as a tactical weapon leaders rely on under pressure. Pete Blaber emphasizes that great leaders control their tone and presence under fire. During high-risk operations in Afghanistan and Iraq, he learned that calmness amid chaos was essential. When plans unraveled and danger escalated, teams took their cue from the leader's voice. Blaber recounts a mission in northern Iraq when his team was operating in close proximity to Kurdish Peshmerga forces. A misstep by the allied fighters caused a premature movement, nearly exposing his Delta team to an enemy ambush. The terrain was unforgiving, and the element of surprise was slipping. But Blaber didn't yell or overreact. He calmly repositioned his team, issued quiet, direct guidance over the radio, and shifted the operation's pace to regain control. His composure prevented a compromised position from becoming a firefight and the mission was completed without a single U.S. casualty. The same applies in shipboard emergencies. In casualty control, combat, or crisis, the most effective leader is the one who mitigates the chaos with calm, steady decisions. Calm doesn't mean slow, it means deliberate. It reassures others that someone is in charge, and that control is possible even in the storm.

Take the Helm

- **Project Calm in Crisis** – Use calm communication as a leadership tool. Practice controlled voice projection during casualty drills and stressful scenarios.

- **Train for Tone** – Coach junior leaders on how to use their presence to stabilize the watch team or crew, especially during surprise evolutions.

Eyes on the Horizon

In the chaos, your calm becomes their confidence. Lead with poise and your team will rise under pressure.

Day 245 – Warfighters Win Through Preparation, Not Posturing

Fearless – Eric Blehm

"Courage isn't bravado. It's humility, training, and relentless effort."

Elite performance doesn't come from bravado, but rather from consistent preparation especially when the path gets harder. Author Eric Blehm tells the story of Navy SEAL Adam Brown, whose journey to elite performance had everything to do with preparation grounded in humility. After losing vision in his dominant eye during a training accident and later crushing the fingers of his shooting hand in a vehicle door, Brown could've stepped back. Instead, he doubled down on preparation. He taught himself to shoot with his non-dominant hand, retrained his body, and mastered his tactics all over again without complaint or fanfare. Blehm shows that Brown prepared for missions, as well as for setbacks. His discipline was defined by how he persistently trained through adversity rather than around it. Whether you're training a damage control party or preparing a ship for INSURV, the best teams aren't defined by what goes right, but how they respond when it doesn't. Determination fuels preparation, and preparation wins the fight.

Take the Helm
- **Grind Before You Shine** – Build repetition and rigor into the most basic routines from watch reliefs to general quarters manning to damage control training team evaluations.

- **Prove Readiness Through Action** – Remind your crew that real readiness is proven in execution, not image.

Eyes on the Horizon
True warfighters prepare. They don't flex. Confidence without preparation is a liability. Grind in peace to win in war.

Day 246 – Lead Through the Fog

The Art of War – Sun Tzu

"When in doubt, make it appear that you know."

Clarity is a weapon even when it's incomplete. Sun Tzu teaches that leaders must project certainty to maintain cohesion, especially when conditions are unclear. War, like leadership, unfolds in chaos. If a leader appears confused or paralyzed by uncertainty, doubt spreads through the ranks. But when a leader steadies the team with composed presence and decisive action whether fully informed or not, they create momentum and control. During the chaotic night battles off Guadalcanal in 1942, U.S. Navy commanders like Admirals Norman Scott and Daniel Callaghan faced Japanese forces in conditions that nullified radar advantage and situational awareness. With darkness masking positions and confusion mounting, they led from the front, issuing orders decisively, maneuvering under fire, and keeping their ships engaged despite the unknown. Losses were heavy, but their aggressive command presence prevented collapse and bought time for the broader campaign to succeed. In moments where communication breaks down, intel is delayed, or conditions change without warning, Sailors look for direction, not perfection. Your ability to act within the commander's intent, apply prior guidance, and move with calm authority keeps the team aligned. You may not know everything, but you must still lead like you know the next step.

Take the Helm

- **Act with Intent** – Rehearse contingency execution in scenarios with degraded visibility or lost comms. Teach Sailors to operate with command intent.

- **Steady the Team** – When leading in ambiguity, give your team clear direction even if it's just the next step. Project control before you fully regain it.

Eyes on the Horizon

Clarity is rare in real-world operations. Project confidence, give guidance, and lead your team through the fog.

Day 247 – Courage Comes in Quiet Moments

Daring Greatly – Brené Brown

"Vulnerability is not weakness. It's our most accurate measure of courage."

Courage in leadership often shows up in the quietest, most personal moments. Brené Brown redefines vulnerability not as a liability, but as the clearest expression of strength. She writes about leaders who admit when they're wrong, ask for help when they're unsure, and open the door for honest conversations, especially when doing so feels uncomfortable. These moments require far more bravery than posturing or pretending to have all the answers. In 2021 and 2022, Sailors aboard *USS George Washington* raised serious concerns about mental health during prolonged shipyard life. While initial leadership responses were slow, a shift began when command triads started engaging Sailors in honest, open conversations, town halls without scripts, questions without judgment, and listening without rushing to respond. It wasn't about having perfect answers. It was about showing up with humility and care. That change didn't make headlines, but it started to rebuild trust where it had been lost. Brown's work shows that leadership rooted in authenticity is more resilient. When leaders dare to be real, they create space for their teams to do the same. And that kind of culture outlasts any checklist or command directive.

Take the Helm
- **Embrace Vulnerability** – Normalize moral courage in your division. Recognize and support Sailors who speak up.

- **Speak the Quiet Truth** – Model quiet strength. Don't shy away from hard conversations or uncomfortable leadership moments.

Eyes on the Horizon
Courage isn't always loud. In leadership, the quietest stands often echo the longest.

Day 248 – Warfighters Know Their Weapons and Their People

Fearless – Eric Blehm

"He trained like a machine, but he led with heart."

Warfighting is a thinking profession built on mastery of both tools and teams. Eric Blehm recounts how Navy SEAL Adam Brown is remembered, not just for his skill with weapons or his tenacity in combat, but for the way he saw and supported the people beside him. Brown trained relentlessly, blind in one eye, missing fingers, operating at the highest level of special warfare, but what set him apart was how well he knew his teammates. He remembered birthdays, checked on families, noticed when someone was off, and backed them without hesitation. That's what true warfighting leadership looks like. It's not just being lethal but being locked in on your team's well-being. Brown didn't separate tactical readiness from relational awareness. He knew both were mission critical. In the Navy, you may know every spec on your weapons system, the radar console, the CIWS mount, the propulsion plant, but if you don't know the Sailor operating it, you're still blind to your unit's full capability. Ships and commands cannot run without people. The leaders who win the hard fights are the ones who invest in both the tools and the team. And your ability to know your team as deeply as you know your gear is what gives you the edge in the fight.

Take the Helm

- **Walk the Deckplates** – Regularly walk spaces with a mindset to observe and understand both the equipment and people too.

- **Build Trust Before You Need It** – Know your Sailors' strengths, weaknesses, and instincts so you can lead them when it's loud and fast.

Eyes on the Horizon
Warfare is about systems, but success is about people. Know your platform but lead your Sailors.

Day 249 – Pressure Reveals Preparation

The 33 Strategies of War – Robert Greene

"In battle, the most prepared often seem the most lucky."

What looks like composure in crisis is often the product of unseen discipline. Robert Greene explains that victory under pressure is rarely improvised. It's built long before the first shot is fired. Leaders who appear unshaken during the storm have rehearsed responses, internalized their strategy, and trained their teams to function reflexively. He points to Alexander the Great at the Battle of Gaugamela, where his vastly outnumbered force defeated King Darius III's massive Persian army. On the surface, Alexander's maneuvers looked like battlefield brilliance. But Greene reveals it was all built on tireless training, battlefield rehearsals, and a deep understanding of how his team would respond under pressure. When Darius launched his feared chariot assault, Alexander's soldiers calmly stepped aside, just as they had drilled, then closed ranks to strike. His decisive cavalry thrust at the Persian center wasn't a gamble. Trust between commander and unit turned a complex maneuver into a synchronized blow. When naval systems fail or casualties erupt, leaders can't rely on luck or charisma. Ships fight the way they train. Reps, not rhetoric, win the day. Naval leaders must adopt this mindset long before combat arrives. Repetitive drills, brutal honesty in debriefs, and systems built to withstand confusion are essential, not extra.

Take the Helm
- **Make Reps Real** – Create high-stress reps in training. Don't just test systems, test mental readiness.

- **Prepare for the Pressure** – Reinforce with your team: pressure doesn't create failure. It reveals whether you trained for success.

Eyes on the Horizon
In high-stress moments, your performance reflects your preparation. Train honestly, or the pressure will show the cracks.

Day 250 – When It's Your Call, Make It

The Dichotomy of Leadership – Jocko Willink and Leif Babin

"There is no one else. You are the leader. Make the call."

Leadership means accepting that, at times, the weight of the decision rests solely on your shoulders. While empowering others is essential, there are moments when leaders must step forward and make the call, decisively and without hesitation. During Operation Red Wings in 2005, SEAL Lieutenant Michael Murphy and his team were ambushed by overwhelming Taliban forces in the mountains of Afghanistan. Surrounded, wounded, and unable to communicate, Murphy understood no one was coming unless he made it happen. Exposing himself to enemy fire, he climbed into the open to get a clear signal and placed the call for reinforcements. He was fatally shot mid-transmission, but that act saved his teammate's life and embodied the purest form of combat leadership: taking the hit to make the call. In the Navy, these moments may not always come under gunfire. Whether navigating during an engineering casualty or responding to an emergent tactical picture, there will be moments when Sailors will look to you. If it's your decision to make, do not wait, make the call with confidence, training, and ownership behind every word. Leadership isn't always about consensus. Sometimes it's about stepping forward because no one else will.

Take the Helm

- **Lead in the Void** – Teach junior leaders that command presence is not tied to rank. Confidence grows when they step forward to make decisions.

- **Clarify Command in the Chaos** – During walkthroughs and combat drills, ask: "Who makes the call?" and ensure someone is always ready to say, "I do."

Eyes on the Horizon

There will be a moment when the weight is on you alone. Make the call. That's leadership under fire.

Day 251 – Confidence Is Built, Not Claimed

Grit – Angela Duckworth

"Our potential is one thing. What we do with it is quite another."

True confidence stays quiet, rooted in the work and experience that justify it. Angela Duckworth explains that the highest performers aren't those with the rawest talent, but those who persist through failure and commit to long-term goals with discipline and heart. In her research at West Point, Duckworth studied cadets during the brutal first seven weeks of training known as Beast Barracks. What predicted who would endure wasn't test scores or athletic accolades, it was grit. Duckworth tells the story of a cadet who failed early fitness assessments and fell behind his peers, but while others quit or coasted, he showed up each day, pushed himself through pain and failure, and absorbed every lesson. By the end of training, he wasn't the fastest, but he had earned the trust of instructors and teammates, not because he started strong but because he refused to give up. That same truth holds in the Navy. Confidence that holds under pressure develops over reps and revisions, and from showing up when quitting would be easier. The Sailors trusted under pressure are rarely the flashiest. Steady effort and proven reliability earn that trust day after day. That's not claimed confidence. That's earned credibility. And it's the kind people follow.

Take the Helm
- **Turn Setbacks Into Fuel** – Encourage your Sailors to view failure as friction that shapes leadership, not as a setback or obstacle.

- **Do the Reps** – When mentoring, highlight the reps behind the performance. Confidence grows not from comfort, but rather from challenge.

Eyes on the Horizon
Confidence that's earned becomes quiet strength. The leaders your team trusts most are forged in struggle, not spotlight.

Day 252 – Own the Outcome Even When It Hurts

Extreme Ownership – Jocko Willink and Leif Babin

"There are no bad teams, only bad leaders."

Leadership begins and ends with ownership. Admiral Husband E. Kimmel, Commander of the U.S. Pacific Fleet during the attack on Pearl Harbor, faced a similar moment. Despite receiving incomplete and fragmented intelligence from Washington, Kimmel accepted responsibility for the fleet's unpreparedness. He didn't point fingers. He wrote in his resignation that accountability "rests here." His career ended, but his act of ownership became one of the earliest examples of flag-level accountability in U.S. naval history. Owning everything in your lane isn't about deflecting blame but about standing up first when the team needs accountability. Whether you're a division officer facing a failed inspection or a CO responding to a crisis, your willingness to own the outcome shapes the culture around you. Your Sailors aren't expecting perfection. They're watching how you respond when things go wrong. Ownership is leadership. And when you model it, they'll follow. Especially when it's hard. Especially when it hurts.

Take the Helm
- **Turn Mistakes Into Momentum** – Use setbacks to model ownership. Don't shield your Sailors from your mistakes. Teach through them.

- **Lead the Fix** – Encourage your division to adopt "my fault, my fix" culture where accountability drives improvement.

Eyes on the Horizon
When leaders own the loss, the team learns to win. Accountability is the cost of command and the root of trust.

Day 253 – Violence of Action, Precision of Thought

The 33 Strategies of War – Robert Greene

"Be fast and overwhelming but only when you know exactly why."

True combat leadership demands a paradox: decisive, overwhelming action guided by calm, calculated thought. Robert Greene emphasizes that boldness alone is never enough. Successful leaders combine aggression with strategic awareness. Nowhere was this balance clearer than in Operation Neptune Spear, the SEAL raid that killed Osama bin Laden in 2011. The mission moved with stunning speed as helicopters flew low into Pakistani airspace and operators breached walls and cleared buildings with exact precision. But that action was only possible because of exhaustive planning, rehearsals, and intelligence assessments. Every room had been simulated. Every contingency considered. When one helicopter crash-landed during insertion, the team didn't pause. They adapted instantly and were able to do so because they were executing a plan designed for uncertainty. In naval leadership, you'll face moments that demand the same, whether it be launch decisions, emergency maneuvers, or casualty responses. The power of your action must be matched by the clarity of your thought. The fastest move rarely matches an effort grounded in purpose and backed by preparation and control. Move when it counts. Move with intent. That's the balance: decisive force when it counts, anchored by precision of thought before it begins.

Take the Helm
- **Don't Rush to Impress** – Show your team that true controlled aggression relies less on speed and more on disciplined readiness.

- **Balance Speed with Thought** – In tactical scenarios, reinforce the balance: act decisively but never without clarity.

Eyes on the Horizon
Violence of action wins battles when it's paired with precision. Train for both, and your decisions will hit with force and focus.

Day 254 – Stay in the Fight

Resilience – Eric Greitens

"Resilience is not about bouncing back. It's about moving forward despite the weight."

In a naval environment, you will face stress that doesn't go away after a good night's sleep. Failed inspections, personal loss, toxic leaders, family strain, or back-to-back deployments will push you to your limit. There will be times when you feel alone or overwhelmed, and the burden of leadership will still be there. But your Sailors are always watching. They don't expect perfection. They need to see that you can keep going, even when it hurts. Because that gives them permission to do the same. Eric Greitens, a former Navy SEAL, wrote *Resilience* as a series of letters to his teammate Zach, another SEAL who returned from combat scarred, disillusioned, and drifting. The battlefield was behind them, but the internal war had only begun. Greitens challenged Zach not with hollow encouragement, but with action. Keep moving. Don't seek to "bounce back." Build something new. Whether you're in uniform or beyond it, the fight never really ends. But neither does your ability to lead. One honest step at a time.

Take the Helm

- **Lead Through the Low Points** – When your team faces setbacks, be transparent about the challenge then set the tone for how you'll push through.

- **Model Forward Motion** – Reinforce that adversity is part of the mission, not a break from it. Stay focused on movement, not perfection.

Eyes on the Horizon
Smooth sailing hides nothing. The real test of leadership comes when you're knocked down and refuse to stop. Stay in the fight.

Day 255 – The Standard Is the Standard Even in the Fight

Sea Power – Admiral James Stavridis

"Discipline is the steel thread that runs through every successful naval command."

During the War of 1812, Captain Isaac Hull's leadership aboard *USS Constitution* proved that disciplined execution sharpens under pressure. As *USS Constitution* closed with *HMS Guerriere*, the ship took early damage and several rigging lines were shot away, limiting maneuverability. Amid the smoke and splintering wood, orders had to be shouted over cannon blasts, and confusion threatened to creep in. But Hull's crew held fast to their training. Gun crews stayed coordinated, reloaded efficiently, and maintained precise timing, even as casualties mounted. What might have become chaos instead became dominance. The crew's ability to execute cleanly in the chaos turned the tide and earned *USS Constitution* her nickname, "Old Ironsides." Calm-sea preparation didn't win that fight. Victory took shape in the turmoil of close combat, held together by discipline until the tide turned. Today, operational tempo, stress, or crisis can tempt leaders to relax expectations. Don't. Standards are your anchor. This applies during sustained operations, watch rotations, and maintenance backlogs. When standards are upheld consistently, even in the fight, they become a source of strength, trust, and performance. Your crew doesn't need new rules when it's hard. They need to see that the ones that mattered yesterday still matter now.

Take the Helm
- **Reinforce the Standard Daily** – In fast-paced or uncertain environments, reinforce the basics. Standards create structure when everything else feels fluid.

- **Be the Example** – Set the tone: "We don't compromise the standard. We adapt around it."

Eyes on the Horizon
The fight doesn't excuse you from the standard. It demands you hold to it. In chaos, consistency becomes your crew's compass.

Day 256 – When You're Tired, Lead Anyway

Make Your Bed – Admiral William H. McRaven

"Sometimes, the simple act of making your bed can give you the strength to do the next thing right."

Fatigue is guaranteed but leadership is still required. As Admiral William H. McRaven describes it, BUD/S training didn't present exhaustion as a passing test but as the very climate that forged resilience. Leaders who stayed the course acted purposefully, even when they didn't feel fully prepared. Whether it was making a bed with perfect corners after a sleepless night or leading a boat crew through freezing surf, success came from discipline in the small things. When McRaven's class was pushed to the brink during surf torture and log carries, it wasn't physical strength that held them together, it was the quiet resolve of teammates who refused to quit, who led simply by enduring. McRaven's lessons apply directly to shipboard life. When operations run long, sleep is short, and the pace doesn't let up, your Sailors look for someone still in the fight. You don't have to be the loudest, you just have to be the most reliable. Show up, stay sharp, and do the small things right. That's what inspires others to push past their own fatigue. You don't lead by pretending you're not tired, you lead by showing that even when you are, the mission still gets your best.

Take the Helm

- **Lead Through Discipline** – In high-fatigue environments, double down on routines and discipline. They create structure when your edge dulls.

- **Set the Watch Tone** – Your posture and presence on the worst day set the bar for everyone watching.

Eyes on the Horizon

Fatigue tests discipline. When you lead through exhaustion, you teach your team how to do the same when it matters most.

Day 257 – No One Fights Alone

Leading With the Heart – Mike Krzyzewski

"You can't win without your team and your team needs to know you'll fight for them."

When Coach K took charge of the USA Men's Basketball Team for the 2008 Olympics, he knew trust wouldn't come from title alone. So, he personally reached out, not just to stars like LeBron James, Kobe Bryant, and Dwayne Wade, but to every player on the roster. He led by serving. By listening. By being present in every rep, conversation, and tough moment. When pressure mounted, the team played for each other because their coach had made it clear: they wouldn't fight alone. That was the difference. That unity was exactly what had been missing four years earlier. In the 2004 Athens Olympics, Team USA, despite a roster of NBA stars, suffered stunning losses to Puerto Rico, Lithuania, and Argentina, ultimately finishing with a disappointing bronze. Analysts pointed to a lack of chemistry, inconsistent effort, and a group of players who hadn't fully bought into each other or the mission. The 2008 Redeem Team succeeded because it was built on connection, shared sacrifice, and trust. You don't lead by barking orders from a safe distance. You lead by stepping into the fire with your Sailors. You don't have to be perfect. You just have to be with them. Because when the fight comes, no one should stand alone. Not when the leader leads with heart.

Take the Helm
- **Lead Shoulder-to-Shoulder** – Be physically and emotionally present during high-stress moments. Stand next to your Sailors, not above them.
- **Build Loyalty Through Presence** – Build team trust before the fight. Support and unity aren't improvised.

Eyes on the Horizon
Your Sailors will fight for the mission when they know you'll fight for them. Lead shoulder-to-shoulder, not from behind.

Day 258 – Set the Tone for the Fight

Man's Search for Meaning – Viktor E. Frankl

"Life is never made unbearable by circumstances, but only by lack of meaning and purpose."

Before the first order is given or the first round fired, great leaders define the purpose behind the mission. Viktor E. Frankl wrote that purpose gives people the strength to endure suffering and in leadership, it gives teams the strength to push through adversity. In 1779, during the Battle of Flamborough Head, Captain John Paul Jones faced overwhelming odds against *HMS Serapis*. His ship was battered, burning, and sinking. When called upon to surrender, Jones famously replied, "I have not yet begun to fight!" Behind that declaration was no bravado, but a deep belief tied to purpose. His refusal to quit gave his exhausted crew the will to press on, and against all odds, they turned the tide. In naval leadership, you'll face moments where conditions deteriorate, fatigue sets in, and confidence in the mission begins to slip. Your Sailors not only need directions, but they also need to believe. When you connect them to a deeper meaning of honor, legacy, and duty, they'll endure more than they thought possible. Set the tone early. Not with noise, but with purpose. Because belief fuels action at the start and sustains it long after motivation fades.

Take the Helm

- **Align Action With Meaning** – Begin major evolutions or shifts in posture with a simple purpose brief. Tie action to mission.

- **Lead with Purpose** – Remind your team regularly why their work matters. Purpose reduces fatigue and sharpens execution.

Eyes on the Horizon
Purpose gives warfighters strength before the pressure ever arrives. Start with sharing your vision and your team will finish with resolve.

Day 259 – Every Sailor Is a Sensor

The Leader's Bookshelf – Admiral James Stavridis and R. Manning Ancell

"The best leaders know how to listen to the entire ship, not just the bridge."

Information is power but only if you're listening. Admiral James Stavridis and R. Manning Ancell underscore that effective leadership starts with humility and ends with curiosity. The best leaders don't just speak well, they listen across all levels. They walk the ship, ask questions, and tune into what their Sailors are seeing, hearing, and sensing. Stavridis draws from decades at sea to highlight how vital it is to tap into the full awareness of a crew not just what's briefed in the wardroom, but what's observed in the spaces no one visits unless something breaks: the unusual vibration near aft steering, the faint smell of ozone in an equipment room, or a pipe sweating in main machinery. Every Sailor is a sensor whether they're on the helm, in aft steering, or inspecting valves on the midwatch. That means leadership must build a culture where every voice matters. When you value input from the most junior Sailor and respond with respect and action, you not only catch problems, but also build trust. Stavridis and Ancell teach that attentive, inclusive leadership often sees danger before it hits and catches opportunity before it passes. Your people are always watching and listening. The question is, are you?

Take the Helm
- **Listen Below Deck** – Empower your Sailors to report what they see and feel. Even if it's outside their lane.

- **Respond With Respect** – Reinforce that observations, concerns, and questions are welcomed not dismissed.

Eyes on the Horizon
The best teams hear more than they speak. Build a culture where every Sailor contributes to true situational awareness.

Day 260 – Lead for the Fight You Can't Yet See

Great by Choice – Jim Collins and Morten T. Hansen

"The signature of greatness is not the ability to predict the future. It's the ability to prepare for it."

Jim Collins and Morten T. Hansen explore why some teams thrive in uncertainty while others fail. The most successful leaders embrace what they call "productive paranoia," meaning an unrelenting commitment to readiness, even when everything seems calm. It's the leader who double-checks watertight integrity before getting underway or who reviews contingency plans during a simple anchoring evolution. Productive paranoia replaces panic with disciplined thinking and proactive planning. It's discipline with foresight. It's preparing for what's likely and for what's devastating if it happens as if the storm could break at any moment. In 2005, *USS San Francisco*, a fast-attack submarine, struck an uncharted undersea seamount at flank speed. The collision was catastrophic, killing one Sailor and injuring many more. But despite the violence of the impact, the crew stabilized the reactor, saved the vessel, and returned to port. Their survival was the result of relentless scenario-based casualty drills: flooding in the engine room, loss of propulsion, electrical fire in maneuvering. It was the procedural discipline and leadership that trained for the fight no one saw coming. You may not know when your gear will fail or when the threat will emerge, but your leadership can make sure the crew is ready when it does. Because readiness is leadership. And greatness is preparation in motion.

Take the Helm
- **Prepare with Intent** – Lead with the future in mind. Ask, "What would this look like in combat?" and adjust the standard accordingly.
- **Rehearse With Purpose** – Elevate your team's mindset from compliance to preparation. Train like "the big day" coming.

Eyes on the Horizon
You can't see every threat but you can lead like it's coming. Preparedness starts with perspective.

287

Day 261 – Warfighters Protect Their People

The 12 Rules for Life – Jordan B. Peterson

"You must care for your people as if they are family. Because when it matters, they are."

Jordan B. Peterson recalls a story of a father struggling through hardship who found strength in a single, anchoring truth: "Because I have children." He uses this to highlight that genuine care for others, especially those who rely on you, often inspires deeper commitment than self-interest ever could. Peterson discusses that when you treat someone's well-being as your responsibility, it alters your priorities. You become less impulsive, more thoughtful, and more willing to endure hardship to protect what matters. This principle resonates deeply in the Navy where long deployments often pull Sailors away from their actual families and toward a crew that becomes one. Division Officers, Chiefs, and Leading Petty Officers become more than supervisors. They're surrogate parents, mentors, and guardians charged with far more than job performance. They monitor sleep cycles, watch for signs of burnout, help navigate personal crises, and stand up for their teams in planning boards and maintenance meetings. The best leaders don't just fight for operational readiness but instead fight for their people's well-being. That watchful stewardship, much like a father's silent resolve, remains content without applause.

Take the Helm
- **Make Protection Part of the Standard** – In every mission or high-tempo event, check your team's well-being not just their posture or PQS.

- **Lead With Care** – Make clear that toughness includes care. Real leaders push hard, but never at the cost of their people.

Eyes on the Horizon
Combat leadership means protecting your team while preparing them for the fight. The best warfighters lead with both grit and grace.

Day 262 – Fight as You Train, Train as You Fight

The War of Art – Steven Pressfield

"The professional arms himself with patience, not the expectation of inspiration."

A true warrior builds readiness not on bursts of adrenaline but through daily, disciplined, purposeful training. Captain Arleigh Burke's destroyer squadron proved this in 1943 at the Battle of Cape St. George. Acting on intelligence that a Japanese convoy would be transiting the area, Burke and his team prepared for a high-risk night engagement. But the exact moment and location weren't guaranteed and in the blackness of night, everything depended on discipline and instinct. When contact came, the U.S. destroyers executed their attack with flawless precision, sinking multiple enemy ships without taking a single hit. They succeeded because they had trained for this moment in relentless detail, rehearsing silent approaches, radar-guided targeting, and coordinated torpedo launches until they became second nature. Their maneuvers were practiced. Their decisions were reflexive. Their victory was earned long before the first torpedo fired. Steven Pressfield reminds us that greatness is built in the mundane, in the reps, the routines, and the relentless refusal to compromise. This means treating every training evolution with the seriousness of the real fight. Your Sailors may not fight with inspiration, but they'll fight with the habits you've built. Every foul weather jacket donned in a drill, every simulated casualty taken seriously, every critique followed through, that's what wins the day when the fire breaks out or the threat becomes real.

Take the Helm

- **Drive Realism in Training** – Ensure your training reflects the conditions of combat. No shortcuts, no disclaimers.

- **Show Up Like It's Game Day** – Reinforce to your team: training is your battlefield. Win there first.

Eyes on the Horizon

The habits you build in peace are the instincts you'll trust in war. Train like it matters because it will.

Day 263 – Know the Mission, Then Own It

The Mission, the Men, and Me – Pete Blaber

"If you don't understand the mission, you don't understand anything."

Mission clarity is the starting point for every effective decision. Pete Blaber drives home the idea that true leadership begins with understanding the deeper intent behind a given task. During his time with Delta Force, Blaber learned that rigid adherence to a plan without grasping its purpose led to failure when conditions changed. But leaders who understood the mission could adapt, improvise, and make real-time calls that still served the greater objective. That kind of ownership only comes when intent is crystal clear. In naval leadership, this principle is just as critical. From the bridge to the mess decks, Sailors must know what the mission is and why it matters because when things go off script, clarity of purpose is what keeps the ship on course. Mission ownership calls for knowing the commander's intent so clearly that you never hesitate to act, no matter what's missing from the plan. Whether you're a department head adjusting watches or a junior Sailor managing an unexpected casualty, if you fully understand the mission, you can lead with confidence.

Take the Helm
- **Make Decisions Aligned With Purpose** – Before any evolution or deployment, ask your team to explain the mission and its priorities. If they can't, reset.

- **Communicate Commander's Intent** – Give junior Sailors command intent not just procedures. Understanding enables autonomy.

Eyes on the Horizon
If your Sailors don't know the mission, they can't own the outcome. Purpose turns compliance into commitment.

Day 264 – Don't Train Them to Follow; Train Them to Take Over

Turn the Ship Around! – L. David Marquet

"We learn best when we have skin in the game. Give control, and you create leaders."

A genuine leader focuses less on control and more on building new leaders to carry the mission forward. L. David Marquet learned this firsthand when he took command of *USS Santa Fe*, a ship known for poor morale and low performance. Rather than barking orders, Marquet shifted to a model of intent-based leadership, training his Sailors not just to execute but to think, decide, and lead. He stopped giving commands and started asking, "What do you intend to do?" This small change had massive ripple effects. Sailors began pre-briefing maintenance plans, qualifying ahead of schedule, challenging outdated protocols, and making sound decisions at their level without waiting for top-down approval. Over time, *Santa Fe* went from worst to first in retention and operational excellence, earning some of the highest inspection scores in the fleet. Its crew produced more future submarine captains than any other during that period. Marquet built leaders who could thrive without him. Train your Sailors not to wait for instructions, but to anticipate, adapt, and take command when needed. Give them ownership. Let them lead small before they lead big because the truest measure of your leadership is what happens when you're not in the room. Train them to take over. Then trust them to do it.

Take the Helm
- **Build Leader Pipelines** – Build bench depth in your team. Don't just qualify Sailors, mentor them for leadership.
- **Teach Ownership, Not Obedience** – Make space for your people to lead under supervision. Correction teaches more than control.

Eyes on the Horizon
The goal is simple: build leaders who build more leaders. The best warfighters train the next crew to take the watch with confidence.

Day 265 – Keep the Standard When No One's Watching

Ego Is the Enemy – Ryan Holiday

"Discipline in the dark is what builds greatness in the light."

True warfighters are built in silence when there's no recognition, no applause, and no external reward. Ryan Holiday makes clear that ego drives us to seek validation, while discipline drives us to do the work regardless. He reminds us that the most consistent performers are anchored by principle, not fueled by attention, and that greatness is formed in the quiet moments where only character remains. Marcus Aurelius, Stoic philosopher and Roman Emperor, lived this principle daily. Amid war, plague, and betrayal, he privately held himself to a rigorous standard every without seeking praise. He led the world's greatest empire not by image, but by internal discipline. On the deckplates, this looks like the midwatch done right, the log filled out with precision even when the spaces are empty, and the drill taken seriously even when no one's grading. Your Sailors may not see every decision, but they'll feel the culture you build. Lead with quiet consistency and maintain the standard, even when no one's watching. Because the moment that matters won't come with fanfare, it will come quietly. And how you've led in the dark will decide how your team performs in the light.

Take the Helm
- **Be Consistent in the Quiet** – Hold the line on excellence even during routine or unsupervised evolutions. Standards don't take breaks.

- **Choose Discipline Over Ego** – Coach your Sailors that real professionalism is internal built on pride, not praise.

Eyes on the Horizon
Integrity holds firm in the quiet moments when no one is there to judge it. Discipline in the dark builds leaders who last.

Day 266 – The Team Is the Weapon

The Five Dysfunctions of a Team – Patrick Lencioni

"If you want to go fast, go alone. If you want to go far, go together."

In combat, individual excellence matters, but it's team cohesion that wins the fight. Patrick Lencioni explains that high performance stems not just from skill but from trust, open dialogue, and a commitment everyone shares. Dysfunction destroys performance in high-stakes environments, but unity transforms potential into power. You see this clearly during replenishment operations in heavy seas. The equipment is precise, but it's the team that makes it work. Line handlers, bridge teams, riggers, safety observers, and conning officers must communicate instantly, anticipate movement, and trust each other completely. When the swells rise and tension spikes, there's no time for ego or guesswork. The evolution only succeeds when the entire team moves as one. Your systems matter, but your team is the real weapon. A perfectly calibrated radar is useless if the watch team isn't aligned. A rehearsed casualty response falters if egos block communication. Lencioni's work reminds us that the time you invest in building connection, shared briefs, integrated drills, and mutual respect directly translate into combat effectiveness. When the team trusts each other, they adapt faster, recover quicker, and hit harder. That's how ships win.

Take the Helm
- **Build Unity Daily** – Prioritize unity in high-stakes teams. Create shared rhythms, language, and mutual trust.

- **Lead Connection, Not Just Correction** – Reinforce that no matter the billet, everyone fights together and wins together.

Eyes on the Horizon
Technology supports the mission, but trust, cohesion, and clarity complete it. Build the team like it's your most vital system.

Day 267 – Win the First Minute

Leadership in War – Andrew Roberts

"The initial tone, set under pressure, often dictates the outcome."

In combat and command alike, the first minute of any crisis is decisive. On the eve of the D-Day invasion, General Dwight D. Eisenhower faced mounting pressure. Forecasts called for rough seas, low visibility, and high winds; marginal weather conditions that barely met the threshold for a safe launch. Delay risked losing the element of surprise and proceeding meant sending over 150,000 troops into uncertain waters. On June 5, 1944, Eisenhower made the call: "OK, let's go." He accepted the risk and quietly drafted a note taking full responsibility should the invasion fail. Later that evening, he visited with paratroopers of the 101st Airborne, offering calm reassurance before they launched into enemy territory. That clarity and presence shown in the very first moments set the emotional tone for one of history's most complex military operations. The troops didn't just follow orders. They followed his steadiness. In naval leadership, the first minute of a casualty, combat alert, or even a personnel crisis matters. Don't scramble. Stay centered. Don't react. Lead. Winning the first minute doesn't mean having all the answers but rather having the means to provide stability when uncertainty hits. Once your Sailors feel your confidence, they'll find their own. And that's when the fight starts to turn.

Take the Helm
- **Set the Tempo Early** – Train yourself and your team to focus on the first 60 seconds. Build habits of composure and clear orders.

- **Win the First Minute** – In critiques, review the opening response. What tone did the team take, and where did it come from?

Eyes on the Horizon
The first minute matters most. Lead with clarity at the start, and your team will carry the mission through.

Day 268 – Close With the Enemy

The Art of War – Sun Tzu

"To subdue the enemy without fighting is the acme of skill."

Superior warfighting is preemptive not reactive. Sun Tzu teaches that the highest form of victory is achieved not through brute force, but through strategic mastery: shaping the battlefield, controlling tempo, and denying the enemy options. Victory goes to the leader who maneuvers before contact, deceives before detection, and influences the outcome before the first shot is fired. At the Battle of Midway, Admiral Chester Nimitz applied that principle with devastating effect. By trusting U.S. cryptologists who broke the Japanese code, he knew their plan before they struck. He positioned his carriers, *USS Enterprise*, *USS Hornet*, and *USS Yorktown*, in ambush. When the Japanese fleet launched their initial wave, they believed they were in control. In reality, the battlefield had already been shaped against them. U.S. dive bombers exploited that advantage, delivering a crushing blow that changed the course of the war. This mindset applies to every tactical watch, presence patrol, or high-stakes decision. Closing with the enemy does not always require force-on-force action. Denying the enemy options before they can act is often more decisive. Position early. Move with intent. And lead like the outcome depends on your preparation. That's how warfighters win without losing control.

Take the Helm

- **Position with Purpose** – Reinforce tactical thinking at all levels. Encourage your team to act early, not react late.

- **Lead With Intent** – Train with initiative-based scenarios. Reward well-reasoned action, not just reaction.

Eyes on the Horizon

The fight is won in positioning, presence, and purpose. Lead your team to shape the battle before it ever begins.

Day 269 – Command Without Ego

Ego Is the Enemy – Ryan Holiday

"Confidence is earned. Arrogance is assumed."

Ryan Holiday warns that arrogance is often a mask for insecurity: brittle, performative, and self-serving. Confidence, by contrast, is grounded in preparation and humility. Captain James Lawrence, commanding *USS Chesapeake* in 1813, faced the seasoned *HMS Shannon* in one of the War of 1812's most intense ship-to-ship engagements. Unlike the posturing common in pre-battle rhetoric of the time, Lawrence didn't inflate expectations or issue grand speeches. His focus was inward on readiness, seamanship, and his crew. Reports show he spent his final hours before the battle walking the deck, encouraging his Sailors, reviewing gunnery stations, and preparing quietly for what he knew would be a difficult fight. When he was mortally wounded just minutes into the engagement, he didn't shout orders for revenge or demand personal recognition. His last command, "Don't give up the ship!" was not about himself. It was a call to purpose, to the mission, and to those who would carry it forward. That legacy endured not because it was loud, but because it was selfless. Holiday reminds us that ego demands the spotlight, but true leadership makes space for others to rise. In crisis, Sailors read more than your words. They read your posture, your tone, and your intent. When you lead with humility, your calm becomes their courage. And when the ego steps aside, trust steps in.

Take the Helm
- **Earn Confidence Daily** – Teach your leaders to let their preparation and results speak louder than their words.

- **Lead With Quiet Clarity** – Stay humble especially after success. Arrogance invites error, while quiet confidence builds trust.

Eyes on the Horizon
Arrogance fades fast under fire. Lead with steady confidence and your team will meet the moment with strength and humility.

Day 270 – Fight for the Fleet, Not Just Your Ship

Sea Power – Admiral James Stavridis

"It's not about the ship. It's about the mission, the region, and the alliance."

Naval warfighters should fight for the fleet, not just their ship. Admiral James Stavridis reminds us that every tactical decision at sea contributes to a broader strategic landscape that ripples far beyond the waterline. He shares how, as a destroyer captain in the Mediterranean, he emphasized regional partnerships over individual ship metrics. Escorting merchant vessels, collaborating with NATO forces, and conducting port calls were strategic moves to build trust, deter aggression, and promote stability. Later, as Supreme Allied Commander Europe, Stavridis pushed beyond national pride to unify 28 NATO nations under one operational vision. He consistently advocated that even a single ship's actions, whether how it maneuvered or what message it signaled, had implications for alliance cohesion and regional deterrence. But gaining that global perspective starts on the deckplates. Junior Sailors must understand how maintaining readiness, executing an escort, or conducting a boarding shapes the maritime environment. The mission is bigger than the hull number. Leaders who instill this awareness produce Sailors who fight smarter, think broader, and act with strategic purpose. Because when each ship operates with the fleet in mind, the Navy transforms from separate commands into a cohesive warfighting force that knows its role across the theater.

Take the Helm
- **Lead Beyond the Lifelines** – Reinforce the big picture during ops and briefs. Tie every mission to theater and fleet impact.

- **Teach Strategic Context** – Encourage your Sailors to learn about joint operations, allied partners, and strategic context.

Eyes on the Horizon
You don't just serve a ship. You serve a fleet, a mission, and a nation. Lead your Sailors to think like warfighters with a global view.

CHAPTER 10:
TACTICAL AND STRATEGIC THINKING

Day 271 – Win Without a Shot

The 33 Strategies of War – Robert Greene

"The supreme art of war is to subdue the enemy without fighting."

Victory doesn't always require violence. Robert Greene shows that the most skillful commanders often win through influence, positioning, and presence, not force. In 2014 and again in 2016, *USS Donald Cook* conducted patrols in the Baltic and Black Seas, regions marked by rising tensions and aggressive Russian posturing. Both times, Russian aircraft conducted dangerously close flybys, attempting to provoke a response. These actions weren't random. They were part of a deliberate campaign to intimidate, test NATO resolve, and bait U.S. forces into overreacting thereby providing Moscow a propaganda win or justification for further escalation. But *USS Donald Cook* held the line. Its crew maintained bearing, radio discipline, and calm professionalism throughout the encounter. The ship didn't flinch. It didn't escalate. It didn't need to. Its presence, posture, and restraint sent a stronger message than any weapon could: the U.S. Navy would not be intimidated or manipulated into conflict on someone else's terms. This principle plays out in every presence operation, freedom of navigation transit, and watch rotation. A firm voice, steady course, and controlled demeanor can speak louder than missiles ever need to. Because when you lead with strength and control, you force adversaries to reconsider the fight before it even begins.

Take the Helm

- **Win With Discipline** – Drill CIC and bridge watchstanders on escalation-of-force protocols, ROE, and hailing procedures. Tone and posture are tactical tools.

- **Master the Art of Presence** – Use area intelligence briefs to give watch teams context during presence ops. Help them understand how posture shapes perception.

Eyes on the Horizon

The warfighter's job is to prevail, not to provoke. Power shown with restraint often speaks louder than weapons ever could.

Day 272 – Prepare to Fight Degraded

Discipline Equals Freedom – Jocko Willink

"You won't always have the perfect tool. That's why you train to win without it."

Combat rarely gives you the luxury of ideal conditions. Jocko Willink stresses that true readiness means being able to operate when everything goes sideways, when the gear fails, the plan breaks, or the situation flips. At the Battle of Savo Island in 1942, Allied cruisers patrolling off Guadalcanal were caught off guard by a Japanese night assault. Despite having radar onboard several ships, the technology was underused or misunderstood. Communication across picket lines was poor, and readiness among crews varied widely. When the Japanese opened fire, confusion reigned. Ships struggled to identify targets, relay bearings, or return fire with coordination. Four Allied cruisers were sunk in a single night, making it the worst defeat in U.S. naval history at sea. But the loss led to a critical shift: the Navy overhauled night-fighting doctrine, emphasized radar training, and drilled crews relentlessly to respond under degraded visibility and chaotic conditions. Readiness wasn't redefined by equipment, but by mindset and preparation. In today's Navy, degraded environments are a certainty, whether from radar loss, comms failure, or system disruption. Your Sailors must know how to act when tools fail because their instincts, not their interfaces, will carry the fight forward when it matters most.

Take the Helm
- **Train for Friction** – Include degraded mode scenarios in CIC and bridge drills. Practice fighting with partial capability.
- **Build Mental Flexibility** – Reinforce manual backups and visual comms with watch teams. Build flexibility into the warfighting rhythm.

Eyes on the Horizon
The fight won't wait for ideal conditions. Lead your crew to perform with what's left and still win.

Day 273 – Elevate From Tactical to Operational

Leadership Strategy and Tactics – Jocko Willink

"Don't get so close to the fight that you can't see the battlefield."

As leaders rise, so must their perspective. Jocko Willink emphasizes the need to step back and see the bigger picture. Tactical leaders focus on immediate tasks, but operational leaders understand how those tasks fit into broader plans, systems, and strategic objectives. You can't lead effectively if you're locked into the weeds. To command well, you must learn when to detach, mentally and physically, and assess the full battlespace. During the Battle of the Atlantic, U.S. and Allied ship commanders escorted convoys through U-boat-infested waters. Individually, each ship's task was tactical: protect the convoy. But the war wasn't won by sinking submarines. It was won by ensuring that troops, fuel, and supplies reached Europe. Captains had to resist the urge to chase submarines or engage recklessly. Their restraint, discipline, and coordination supported the operational goal of sustaining the Allied war effort. Today, your section's maintenance, your watch team's coordination, your ship's patrol all contribute to regional stability and strategic posture. The strongest leaders know when to zoom out and align their actions with the broader mission. Because in war, tactical brilliance only matters if it supports operational success. Lead with that perspective, and you lead with impact.

Take the Helm

*   **Think Two Levels Up** – During planning boards or CIC briefs, ensure watch teams understand how your platform supports broader operational phases. Integrate ESG/CSG goals into your comms.

*   **Step Back to See More** – Encourage OODs and TAOs to "zoom out" during watches. Read OPTASKs, analyze area tasking, and understand how ship movement supports theater strategy.

Eyes on the Horizon
Tactical leadership wins the hour. Strategic thinking wins the campaign. Lead with eyes that see beyond your immediate lane.

Day 274 – Anticipate the Second and Third Order

Great by Choice – Jim Collins and Morten T. Hansen

"The signature of great leaders is productive paranoia, thinking ahead without paralysis."

Every decision sets off a chain reaction. Elite leaders not only focus on first-order effects, but they anticipate the second and third as well. This mindset drives leaders to think critically about downstream consequences, such as tempo shifts or logistics strain, without becoming paralyzed by possibilities. In Operation Earnest Will, the United States made the tactical decision to reflag Kuwaiti oil tankers during the Iran-Iraq War, placing them under American naval protection. While the initial move was diplomatic, the second-order effects escalated rapidly: U.S. warships became targets, which resulted in the *USS Samuel B. Roberts* striking a mine. The third-order consequence, Operation Praying Mantis, the largest surface naval engagement for the U.S. since World War II, was a significant escalation resulting in the destruction of Iranian naval vessels. While tactically successful, it marked a sharp increase in hostilities and regional tension. These outcomes were foreseeable by those willing to think beyond the immediate move. In every evolution, ask: What happens if we're right? What happens if we're wrong? A simple escort mission or course adjustment may seem isolated, but it can alter operational tempo, regional posture, and even diplomatic balance. The best leaders go beyond planning the next move and anticipate the chain, because in warfighting, small decisions often ripple farthest. Stay sharp. Think ahead. Lead deeper.

Take the Helm
- **Think in Layers** – During planning, war games, or scenario drills, have your team map out second and third order effects of every major decision.
- **Pause for Impact** – In critique, go beyond the outcome. Review what decisions unintentionally shaped future friction points.

Eyes on the Horizon
When you think beyond the first move, your decisions shape the momentum that follows. Success becomes a product of foresight, not chance.

Day 275 – Simulate to Dominate

The Toyota Way – Jeffrey K. Liker

"Without standards, there can be no improvement. Without simulation, there can be no mastery."

Mastery is not built on theory, but instead forged in realistic, high-pressure practice. Jeffrey K. Liker emphasizes that world-class performance comes from repetition under real conditions, not controlled environments. That mindset is operationalized in the surface Navy through Composite Unit Training Exercises, otherwise known as COMPTUEX, a multi-week, fleet-level event that push strike groups to the edge. During COMPTUEX, every unit, from aircraft carriers to destroyers, faces relentless scenarios: cyberattacks, missile strikes, mechanical casualties, and degraded communications. Fatigue sets in. Coordination is tested. Systems fail. And that's the point. These simulations aren't about passing a test. Every scenario is a chance to sharpen judgment, preparing teams for real-world chaos and the uncertainty and urgency of real-world missions. Watch teams learn to fight through friction. Bridge teams rehearse high-speed maneuvering under stress. Strike leads manage combat airspace under time pressure. Success doesn't hinge on perfection but on consistent readiness. You can't expect flawless performance in crisis if your crew only trains for checklist conditions. Leaders who push their teams through these uncomfortable drills build instincts that hold when plans break. If you want your Sailors to dominate in the fight, you have to simulate the fight every exhausting, foggy second of it. That's how mastery is built.

Take the Helm
- **Make Drills a Rehearsal** – Build stress into your drills. Add equipment failures, crew shortfalls, or late injects to mimic real-world conditions.

- **Stress the System to Strengthen It** – Run CIC and Bridge cross-watch simulations with rotations and fatigue factors. Train adaptability, not perfection.

Eyes on the Horizon
Simulation shapes execution. Train for the fight you expect and prepare for the one you don't.

Day 276 – Align the Bridge and the Fight

The Mission, the Men, and Me – Pete Blaber

"Good units operate in sync. Great ones think in sync."

Pete Blaber emphasizes that elite teams not only execute the same plan but share the same mental model. Great units align intent and decision-making before the action starts. Tactical precision means little if the minds behind it are misaligned. This is especially critical between the bridge and CIC. One owns the ship's position while the other owns the tactical picture, but both shape the outcome. During the 2023–2024 Red Sea anti-air warfare operations, U.S. Navy ships were tasked with intercepting complex, real-time threats including drones and anti-ship missiles launched toward commercial and coalition vessels. Successful engagements depended not only on weapons systems, but on seamless coordination between bridge maneuvering and CIC tracking. Maintaining radar coverage, managing line-of-sight for intercept windows, and ensuring firing arcs remained clear required near-instantaneous communication and mutual understanding between the OOD and TAO. There was no time for clarification, only execution based on shared intent. Strategic leaders ensure these spaces aren't just linked by the net, but by synchronized thinking. That's what turns a ship into a warfighting platform and not just a collection of watchstanders. When the OOD and TAO speak the same operational language, anticipate each other's actions, and understand each other's priorities, the ship moves and fights as one.

Take the Helm
- **Foster Cognitive Sync** – Build shared briefing routines between bridge and CIC. Use common threat models and maneuver plans.

- **Prioritize Intent Over Scripts** – Train your teams to understand *why* decisions are made, not just what to do so they can act without waiting for a call.

Eyes on the Horizon
Seamless operations begin with shared understanding. Align the voices of the ship, and it will move as one.

Day 277 – Align Tactical Action With Strategic Intent

Start With Why – Simon Sinek

"When people understand the purpose behind their actions, they perform with conviction."

Great leaders connect the small things to the big picture. Simon Sinek reminds us that people commit more deeply when they understand the purpose behind their actions. When Sailors grasp the strategic *why*, their routine tasks gain meaning, and their focus sharpens. During the Cuban Missile Crisis in 1962, U.S. Navy ships executed a naval quarantine to block Soviet missile shipments. While the strategic goal was to avoid nuclear war, that success depended on extreme tactical discipline. Every intercept, every radio call, every maneuver had to be intentional, professional, and aligned with Presidential intent. A single misstep at sea could have triggered global escalation. It didn't because tactical actions served a clearly defined strategic purpose. That's the lesson for leaders today: your Sailors must know that what they log, report, or repair isn't just a box to check, but instead a contribution to a larger mission. Train them to connect precision with purpose because when tactical execution is aligned with strategic clarity, every action strengthens the mission and moves it forward.

Take the Helm

- **Tie Tasks to Purpose** – During watch turnover, connect tactical responsibilities to the broader mission. Brief why it matters, not just what to do.

- **Reinforce the Why** – Reinforce with your team that precision in the small things, logs, reports, readiness, protects strategic integrity.

Eyes on the Horizon
The mission lives in the details. Every tactical act should echo with purpose and every leader should make that clear.

Day 278 – Command the Narrative

The Leader's Bookshelf – Admiral James Stavridis and R. Manning Ancell

"Leaders don't just execute the mission. They shape how the mission is understood."

Controlling the message is as vital as controlling the fight. Leaders must communicate clearly, not just to their teams, but to allies, partners, and the public. Intent, values, and action must be aligned and well-articulated. The 1964 Gulf of Tonkin incident underscores the danger of failing to command the narrative. Under poor visibility and high seas, radar misinterpretations aboard *USS Maddox* and *USS Turner Joy* led to reports of a second attack by North Vietnamese torpedo boats that likely never occurred. Despite conflicting accounts, the reports quickly reached Washington. President Lyndon B. Johnson, under pressure to demonstrate resolve, used the incident to request broad military authority from Congress and within days, the Tonkin Gulf Resolution passed overwhelmingly. It marked the beginning of large-scale U.S. military involvement in Vietnam, ultimately escalating into a full conflict that lasted over a decade. What began as ambiguous radar signals became the pretext for a war. There was no deliberate deception on the part of the ships' crews but in the absence of narrative control and clarifying leadership, the ambiguity metastasized into strategic action. The lesson is lasting: when leaders don't take charge of the story, someone else will and it may not serve the truth or the mission.

Take the Helm
- **Communicate With Clarity** – During high-visibility events, review message alignment. Ensure CO, TAO, CDO, and watchstanders are speaking with one voice.

- **Own the Narrative** – In moments of uncertainty or friction, step up to lead the story, not just the strategy.

Eyes on the Horizon
In complex ops, how you explain the mission is part of the mission. Lead the narrative or someone else will.

Day 279 – Be Decisive in the Gray

The Unforgiving Minute – Craig Mullaney

"You won't always have enough. Enough time, enough information, enough clarity. Lead anyway."

Craig Mullaney recounts how real leadership emerges when you're forced to act with incomplete data, under stress, and in high stakes. He learned in combat that hesitation was a decision in itself, and often the wrong one. The most effective leaders accept the fog, assess quickly, and act decisively. Far from reckless, it reflects disciplined decision-making in demanding moments. In 1944, at the Battle of Surigao Strait, U.S. naval forces under Rear Admiral Jesse Oldendorf faced a complex situation. Japanese forces were approaching through the narrow strait at night. Exact enemy strength and timing were unclear. Rather than wait for confirmation, Oldendorf positioned his battleships to "cross the T," maximizing firepower at the decisive point of contact. The result was a crushing victory, the last battleship-versus-battleship engagement in history. That's what decisiveness in the gray looks like and moments happen often in the Navy. Unknown radar contacts, ambiguous ROE scenarios, or technical casualties with limited diagnostics, you'll need to trust your preparation, make a clear call, and then give your team something to rally behind. When you make decisions in the gray with calm confidence, your crew gains trust in both your judgment and their own. That trust becomes the real anchor when the picture stays murky and time runs short.

Take the Helm

- **Normalize Imperfect Decisions** – Include uncertainty scenarios in CIC and bridge drills. Build the habit of assessing and deciding with incomplete information.

- **Lead With Informed Confidence** – When clarity is limited, fall back on training, commander's intent, and sound reasoning.

Eyes on the Horizon

Perfect clarity is rare. Your leadership is measured by what you do when certainty is not an option.

Day 280 – Make the Mission Scalable

Extreme Ownership – Jocko Willink and Leif Babin

"If your plan only works with you in the room, it's not a plan. It's a bottleneck."

Jocko Willink and Leif Babin teach that plans must be scalable, repeatable, and executable at every level. If the mission hinges on your constant presence, then it's vulnerable because real leadership means building systems that function without you. In the Navy, scalability allows the fleet to maintain persistent presence with limited platforms and personnel, a necessity in an era of global commitments and constrained resources. The Littoral Combat Ship (LCS) dual-crew model tested this principle, rotating Blue and Gold crews through a single platform to maximize forward presence and time on station. Scalable leadership here meant more ships forward-deployed without exhausting crews, achieving greater coverage and strategic deterrence without proportionally increasing the fleet size. In theory, it was efficient. In practice, however, inconsistent execution revealed gaps. Some crews struggled with incomplete material turnover, uneven qualifications, or diverging cultures. Without constant leadership alignment, accountability blurred. Others encountered friction, reduced readiness, and morale challenges. But commands with strong leadership pipelines and procedural rigor succeeded. The model showed that scaling a mission isn't just about structure, but also about preparation. Success depends on shared standards, empowered junior leaders, and systems that maintain clarity without constant oversight. When done right, the mission continues anywhere, with anyone, because the culture, not just the presence, is what endures.

Take the Helm
- **Build Systems, Not Silos** – Develop watchbills, qual pipelines, and readiness routines that run smoothly without triad supervision.
- **Train Them to Take the Watch** – Empower EDOs, OODs, and LPOs to run evolutions with clarity and accountability then coach them afterward.

Eyes on the Horizon
Leadership isn't about being in every space but rather building a team that leads when you're not there. Structure creates trust that scales.

Day 281 – Synchronize to Maximize Impact

The 5 Levels of Leadership – John C. Maxwell

"Alignment amplifies effort. When a team moves together, momentum follows."

Execution is important but synchronization creates impact. John C. Maxwell explains that great leaders elevate performance by aligning people, priorities, and actions. When everyone understands their role and how it connects to the larger mission, momentum builds. In April 2017, U.S. Navy destroyers *USS Porter* and *USS Ross* launched 59 Tomahawk cruise missiles at Shayrat Airbase in Syria in response to a chemical weapons attack. What made the strike effective wasn't firepower alone, but the tight synchronization behind it. The operation required real-time intelligence validation, coordination with regional allies, and precise timing to ensure minimal collateral damage while maximizing strategic effect. Communications spanned across combatant commands, fleet assets, and national command authorities. Navigation teams, combat systems, and missile crews had to execute flawlessly within a narrow launch window. Every second mattered. The strike significantly degraded Syria's ability to deliver chemical weapons, demonstrated U.S. resolve, and sent a clear deterrent message without escalating to broader conflict. It also reinforced allied confidence in U.S. naval precision and readiness. Synchronization of intent, timing, and execution transformed a tactical action into a strategic statement. Onboard your ship, the same logic applies. Replenishments, launches, evolutions, they all require unity of effort across departments. Momentum is built when the entire team moves as one, with timing, purpose, and precision. Synchronization is unity of effort. And that unity wins.

Take the Helm

- **Align Before You Execute** – During complex evolutions, brief "sequencing expectations" clearly. Show how each department's timing impacts the whole.

- **Think Beyond Your Space** – Conduct after-action reviews with cross-department feedback. Create a shared picture of performance.

Eyes on the Horizon

Tactical brilliance means nothing without coordination. Unity of effort turns execution into excellence.

Day 282 – Understand the Weapons and the War

Fearless – Eric Blehm

"Mastery is built from obsession with every tool you might one day depend on."

Navy SEAL Adam Brown trained with relentless intensity, mastering his gear down to the smallest detail. After losing vision in one eye, Brown learned to shoot with his non-dominant hand, persistently rehearsed contingency plans, and studied mission terrain in detail. He constantly prepared for how his gear would perform under real conditions, in darkness and in chaos. During the 2010 assault on an enemy compound in Afghanistan's Komar Province, Brown was shot in both legs but continued the mission. His obsessive familiarity with his gear meant he could operate even while wounded, adapt, and execute his role instinctively in the most degraded conditions. His team didn't have to compensate for him. In fact, they were empowered by his preparation. Despite injuries, he provided cover for his team, maintained control of his weapon, and exposed himself to enemy fire to protect others. His actions enabled the SEALs to complete the objective, neutralize the threat, and return safely. Brown was killed in action, but his preparation and understanding of his weapons saved lives. It's not enough to know how systems work. Sailors must understand how that system integrates into real-world scenarios, amid threat timelines, and in degraded conditions. Leaders ensure their teams connect the dots between platform and purpose. When your Sailors know how to employ the weapon and why it matters in the context of the mission, they move from qualified to ready.

Take the Helm

- **Bridge Skill and Judgment** – Integrate tactical employment scenarios into system training. Teach your teams how, when, and why to employ weapons.

- **Tie Training to Threats** – Include brief vignettes from real-world ops in combat system training. Context drives performance.

Eyes on the Horizon

Mastery is more than maintenance. Know your systems and know the war they're built for.

Day 283 – Strategic Risk Is Still Risk

The Dichotomy of Leadership – Jocko Willink and Leif Babin

"Being aggressive does not mean being reckless. Balance boldness with judgment."

Leadership requires risk but not recklessness. In 2009, Captain Richard Phillips was taken hostage by Somali pirates after offering himself in exchange for the safety of his crew from the *Maersk Alabama*. The Navy's response was fast but measured. *USS Bainbridge* established a perimeter, de-escalated immediate tension, and waited for the right moment. Meanwhile, SEAL Team snipers were covertly inserted under cover of darkness. The situation was volatile: the lifeboat was mobile, the pirates were armed, and the distance to shore and potential reinforcements, was shrinking. Rather than rush the rescue, Navy leadership exercised patience and discipline. Only when the risk to Captain Phillips reached a critical threshold and the conditions allowed for a clean shot did they authorize simultaneous engagement. Three snipers fired in unison, killing the pirates and rescuing Phillips without further harm. This operation demonstrated that strategic risk demands judgment. The SEALs did not rely on adrenaline alone. Authority and precision guided every move when the stakes were highest. Being bold doesn't mean moving fast. It means knowing exactly when to move, why, and with what consequences. That's what separates warfighters from gamblers.

Take the Helm

- **Own the Hard Call** – Don't let pressure override your responsibility to assess and act in the long-term interest of the mission and crew.

- **Exercise Bold Judgment** – Be aggressive in execution but deliberate in deciding when the risk outweighs the reward.

Eyes on the Horizon

Strategic risk requires strategic responsibility. Take bold action but lead with disciplined awareness.

Day 284 – Think Beyond the Battle

Call Sign Chaos – Jim Mattis & Bing West

"Tactical excellence wins battles. Strategic vision wins wars."

In the South China Sea, U.S. Navy ships regularly conduct Freedom of Navigation Operations (FONOPs) which are routine, lawful transits through contested waters. These movements may seem procedural, but they're far from routine in their purpose. Each patrol is a deliberate act of strategic signaling: a visible, consistent reminder that no nation has the right to unilaterally rewrite international law. If left unchallenged, excessive maritime claims could become de facto norms, reshaping global sea access and giving authoritarian powers outsized control over vital trade routes. FONOPs reinforce the long game. They uphold the rules-based order and preserve access for all, not just today, but for generations of mariners to come. The ship may not fire a shot, but its transit defends the principle that no one nation owns the sea. This is strategic leadership in action. A single bridge-to-bridge call, a tightly executed maneuver, or a professional radio report contributes to a broader campaign of influence and deterrence. Train your Sailors not just to do things right but to understand why it matters at scale. Tactical wins are necessary but only if they reinforce long-term stability and mission credibility. The real victory lies not in the battle but in shaping the environment that makes the next one less likely.

Take the Helm
- **Pause Before You Press** – In high-stakes moments, take a breath, scan the full picture, and confirm alignment with mission intent before acting.

- **Train Strategic Awareness** – Train your TAOs, OODs, and watch teams to think in terms of commander's intent. If comms drop, they should still know what success looks like at the operational level.

Eyes on the Horizon
Tactics win moments. Strategy wins missions. Lead your Sailors to see beyond the immediate fight.

Day 285 – Lead With Left and Right Limits

Crucial Conversations – Kerry Patterson, Joseph Grenny, Ron McMillan, Al Switzler

"Boundaries don't restrict action. They focus it."

In *Crucial Conversations*, the authors explain that high-stakes situations demand psychological safety and clear expectations, and that people are far more likely to act decisively when they know the boundaries of their authority. Rather than micromanaging every move, strategic leaders establish left and right limits that allow their teams to move fast without crossing the line. The goal is clarity that empowers judgment under stress. That clarity often comes through Rules of Engagement (ROE), commander's intent, and theater-specific doctrine. ROE are legal and operational directives that outline when, where, and how force may be used. Commander's intent explains the purpose behind the mission, guiding action even when plans break. Theater doctrine tailors those expectations to regional threats, norms, and escalation risks. Your Sailors may have seconds to make a call and they won't always have time to reach back for guidance. But when they've been trained to understand what's authorized, what's expected, and what's off-limits, they can act with calm authority. And the leader's job is to make those limits unmistakable before the moment comes. Left and right limits don't just protect the mission, but give your people the freedom to lead boldly, knowing where the boundaries are and how to maneuver within them.

Take the Helm
- **Define the Edges** – Conduct left/right limit drills with CIC and bridge teams. Walk through unclear scenarios and reinforce what's allowed.

- **Train for Decision Space** – Use scenario-based walk-throughs to build judgment within those limits, especially for watchstanders and key leaders.

Eyes on the Horizon
Freedom of action depends on boundaries. Set them early, reinforce them often, and your team will lead decisively.

Day 286 – Train the Ship to Think

Mindset – Carol S. Dweck

"People with a growth mindset thrive on challenge and stretch themselves beyond limits."

A ship full of checklists is not a thinking ship. Carol S. Dweck introduces the power of a growth mindset, a belief that abilities and intelligence can be developed through effort, feedback, and persistence. Leaders with this mindset train for compliance and curiosity. The Navy's evolution toward the Composite Warfare Commander (CWC) model exemplifies this mindset in action. CWC decentralizes tactical authority across various warfare commanders like air, surface, subsurface, and strike, so that decisions can be made quickly by those closest to the fight. This structure only works when every player understands the mission, questions assumptions, and thinks critically under pressure. It demands communication, trust, and initiative, not rote obedience. It turns a strike group from a collection of ships into an intelligent, adaptive warfighting team. When leaders empower their teams to innovate within bounds and improve the process, not just execute it, the ship becomes faster, smarter, and more resilient. You're not just training operators. Every decision, conversation, and empowered Sailor shapes a thinking organization.

Take the Helm
- **Empower Tactical Creativity** – Let your Sailors contribute to procedures, checklists, and solutions. They'll own what they build.

- **Build a Growth Culture** – Run evolution debriefs where junior leaders lead the reflection. Grow thinking, not just compliance.

Eyes on the Horizon
Great leaders go beyond muscle memory. Building mental agility and rapid adaptation are what set teams apart. Train the ship to think then trust it to act.

Day 287 – Know When to Shift Gears

Great by Choice – Jim Collins and Morten T. Hansen

"Leaders know when to push and when to pause."

Greatness comes from disciplined pacing, not constant acceleration. Jim Collins and Morten T. Hansen introduce the "20 Mile March" concept. They found that companies that endured chaos and outperformed peers didn't sprint when times were good or collapse under pressure. Clear performance boundaries, upheld with resolve and restraint, made the difference. They write, "Fanatic discipline is consistency of action aligned with goals, values, and performance standards over time." Strategic leaders apply this same mindset to operational tempo. It's tempting to push hard until the wheels come off, but endurance wins the mission. Collins and Hansen highlight leaders who knew when to throttle up and when to stabilize, even when competitors burned themselves out. That judgment, to surge when it counts and pause when it protects long-term readiness, is what separates good commands from great ones. True leadership means harnessing momentum and keeping it aligned with purpose. The "20 Mile March" exemplifies this by forcing leaders to maintain progress under pressure without overreaching, a mindset that sustains morale, mitigates burnout, and keeps your team mission-capable when others are spent. That kind of discipline in the Navy defines operational excellence over the long haul.

Take the Helm

- **Read the Signs** – Monitor crew energy and team tempo like you monitor systems and know when to ease off to preserve long-term performance.

- **Set Sustainable Pace** – Reinforce that slowing down to recover is a form of leadership, not weakness. Model balance.

Eyes on the Horizon
You don't win by sprinting every mile. Pace your team to last and they'll meet the mission with strength to spare.

Day 288 – Lead With a Bias for Action

Good to Great – Jim Collins

"A culture of discipline is not a principle of business. It is a principle of greatness."

Great organizations create momentum through decisive, consistent action. Jim Collins introduces the "Flywheel Effect," the idea that breakthrough success isn't the result of one dramatic event but the accumulated impact of disciplined effort, day after day. At first glance, Captain David McCampbell's legendary sortie during the Battle of Leyte Gulf, where he singlehandedly downed nine enemy aircraft, seems like a singular act of heroism. But that moment was possible only because of the rigorous training, preparation, and muscle memory developed over years of disciplined flying. McCampbell didn't just act bravely in the moment. He was the product of a command culture that emphasized constant readiness and unwavering commitment to excellence. His ability to take immediate initiative under pressure wasn't luck but rather was the natural result of a flywheel that had already been spinning. This is precisely the power of disciplined action over time: when the stakes are high, there's no need to scramble for brilliance. You fall back on the momentum you've built. In naval leadership, that means cultivating a climate where readiness is routine, excellence is expected, and decisions are backed by preparation. When you lead with a bias for action grounded in discipline, you build greatness, one deliberate step at a time.

Take the Helm
- **Build Decisive Culture** – Rehearse decision-making drills where junior leaders must act without seeking permission. Build trust through action.

- **Execute With Discipline** – Reinforce that action isn't about speed alone, but about purpose, alignment, and confidence.

Eyes on the Horizon
Leaders don't wait to be told. The moment calls, and leadership responds. Action drives momentum, and momentum wins fights.

Day 289 – Bridge Tactical Discipline With Strategic Flexibility

The Art of War – Sun Tzu

"Water shapes its course according to the nature of the ground over which it flows."

Sun Tzu teaches that supreme commanders succeed not by rigidly adhering to plans, but by aligning unwavering principles with adaptive execution. He writes, "The skillful warrior can modify his tactics in relation to the opponent and thereby succeed." This philosophy is tested daily in places like the Strait of Hormuz, where U.S. Navy ships must execute high-stakes transits through contested waters. In one 2019 incident, *USS Boxer* transited the strait under tight ROE and pre-planned formations when it was approached by an Iranian UAV flying aggressively close. Rather than escalate through force posture or rapid maneuver, the crew employed a non-kinetic countermeasure using electronic jamming to down the drone without firing a shot. The decision upheld commander's intent, avoided provocation, and maintained freedom of navigation without breaching diplomatic thresholds, adapting in real time to a dynamic threat. This is the essence of strategic flexibility: bridge and CIC teams don't improvise. Maneuvering happens with purpose, guided by clear boundaries and an awareness of shifting conditions. When leaders train their forces in precise fundamentals but allow for responsive execution within the framework, the result is operational harmony. Fluid, deliberate, and resilient under pressure.

Take the Helm

- **Train Responsiveness** – Build SOPs with flexibility in mind. Train your teams to adapt, not abandon, when friction appears.

- **Lead with Adaptable Structure** – Include "adapt and overcome" vignettes in briefs. Show your Sailors how flexibility fits within warfighting discipline.

Eyes on the Horizon

Discipline anchors the team. Flexibility moves it forward. Strategic leaders balance both with precision.

Day 290 – Train for Strategic Endurance

Man's Search for Meaning – Viktor E. Frankl

"What is to give light must endure burning."

Endurance is purpose sustained through adversity, not merely survival. Viktor E. Frankl, writing from the crucible of a concentration camp, teaches that those who endured the longest were not the strongest physically, but those who anchored their suffering to a greater meaning. He writes, "Life is never made unbearable by circumstances but only by lack of meaning and purpose." For leaders, this insight is transformative: to sustain your team through hardship, you must lead with purpose, not just plans. After the attacks on 9/11, *USS Carl Vinson*, already deployed, was ordered to remain on station and not return home. Instead, the crew surged into the Arabian Sea and became the first naval force to launch combat operations in Afghanistan. The timeline was indefinite. Conditions were tense. But driven by the knowledge that they were delivering the nation's immediate response to an unprecedented tragedy, the crew leaned in. That clarity of purpose aligns perfectly with Frankl's insight: when people understand why they are enduring, they can endure anything. Strategic leaders manage exhaustion and also give it meaning. When your Sailors believe in the mission, they survive, and they sustain.

Take the Helm

- **Plan for the Long Game** – During long ops, plan endurance from the start. Rotate responsibility, protect rest, and watch for burnout signals.

- **Pace the Mission** – Coach your Sailors to balance performance with preservation. Strategic readiness depends on resilience.

Eyes on the Horizon

Leadership is about longevity, not intensity. Pace your team with transparency, and they'll carry the mission farther than expected.

Day 291 – Think in Campaigns, Not Events

The 33 Strategies of War – Robert Greene

"Do not fight isolated battles. Fight the campaign."

True strategic leadership requires a shift in mindset, from short-term reaction to long-term design. In his book, Robert Greene reminds us, "The strategist must see the whole chessboard." One of Greene's most compelling insights is that overreaction to every provocation weakens a leader's position. He highlights the failures of reactive generals and contrasts them with those who endured losses without losing vision, including General George Washington, who often avoided direct battle to protect the broader revolutionary cause. Greene emphasizes that Washington's brilliance wasn't in defeating the British in open-field combat, but in strategically preserving the Continental Army's strength, building alliances, and choosing time and place carefully. His retreats, patience, and selective engagements reflected a leader focused not on the next win, but on ultimate victory. Greene calls this "The Grand Strategy," the ability to absorb hits, hold the long view, and build momentum over time. Strategic naval leaders avoid chasing every issue tactically. They align actions to long-term goals, preserve combat power, build readiness, and win across deployments, not just daily battles. "Power is not only the ability to strike," Greene writes, "but also the wisdom to wait."

Take the Helm

- **Lead the Long Game** – Set long-term objectives for team performance. Don't let daily friction pull you off mission.

- **Zoom Out to See the Campaign** – Resist the urge to overcorrect after every setback. Lead with long-term clarity.

Eyes on the Horizon
React to events but lead the campaign. The strongest leaders stay focused on the long win.

Day 292 – Tactical Patience, Strategic Urgency

Leadership in War – Andrew Roberts

"The best leaders wait for the decisive moment and strike with everything."

The most effective leaders don't rush into action but instead position themselves to strike when the conditions ensure maximum impact. Andrew Roberts highlights this through three distinct leadership approaches. Napoleon Bonaparte moved swiftly when necessary, but only after securing the advantage on terrain, morale, and timing. He didn't seek to be the first to act, but the most decisive, demonstrating that power lies in choosing the moment, not just seizing it. Winston Churchill, by contrast, believed in constant momentum, pressing the war effort forward with relentless energy and public resolve. Yet even he recognized the wisdom of tactical patience when he backed General Dwight D. Eisenhower's insistence on delaying D-Day. Eisenhower refused to gamble with suboptimal tides, moonlight, or weather. He waited, not out of caution, but because he understood that success required overwhelming conditions, not just readiness. Together, these leaders teach that patience is not the absence of action, but the discipline to shape the battlefield before committing force. In naval operations, the same principle applies. Strike groups don't launch on first contact, but instead maneuver, assess, and align resources before delivering precision impact. Tactical patience builds the conditions. Strategic urgency delivers the result.

Take the Helm
- **Train for Timing** – Train your team to recognize the difference between speed and timing. Reward smart delay, not just fast execution.
- **Delay With Purpose** – During rehearsals, simulate shifting conditions. Build comfort in pausing for precision.

Eyes on the Horizon
Strike fast but only when it's time. Patience sets the stage for decisive success.

Day 293 – Develop Strategic Communicators

Start With Why – Simon Sinek

"When people know the purpose, they don't need to be told what to do."

The most effective leaders don't just speak, but rather they make themselves understood. Simon Sinek emphasizes that communication gains power when it is intentional, consistent, and rooted in clarity. In high-stakes environments, the goal isn't just to pass information but to ensure it lands with meaning. During the 2021 Freedom of Navigation Operations (FONOPs) in the Black Sea, U.S. Navy warships consistently broadcasted their movements and intentions in accordance with international law. These weren't routine transmissions. Each one served as deliberate signals to friend and foe alike, affirming resolve while minimizing miscalculation. The clarity of message, professional, calm, and anchored in legitimacy, became a strategic asset in a contested region. Every Sailor, whether writing a LOGREQ, issuing a bridge-to-bridge call, or speaking in a combat team debrief, contributes to mission clarity. Strategic leaders train their teams not only in communication procedures, but in delivering messages that are precise, aligned, and purposeful under pressure. Because in complex environments, clarity builds credibility and credibility builds peace.

Take the Helm

- **Speak With Strategic Clarity** – Teach your Sailors to speak with mission focus. Don't just report data, convey intent.

- **Build Communicators, Not Repeaters** – Have junior leaders practice briefs and hails that tie tactical updates to strategic posture.

Eyes on the Horizon

Strategic communication starts with clear purpose. When your message is understood at every level, your mission moves forward with unity.

Day 294 – Simplify the Complex

The Toyota Way – Jeffrey K. Liker

"Complexity hides failure. Simplicity reveals readiness."

The hallmark of great leadership is how clearly you communicate what matters. Jeffrey K. Liker emphasizes that the most effective systems are simple, visual, and repeatable. He writes, "Standardized tasks are the foundation for continuous improvement and employee empowerment." In complex environments, confusion breeds errors. Toyota's strength came not from oversimplifying but from making complexity manageable. Liker describes how lean systems break down processes into visual, understandable components that empower workers to own quality and anticipate problems. In military terms, this is command and control at its best: decentralizing execution through shared understanding. That same approach proved effective when the Navy implemented Lean Six Sigma initiatives across maintenance and supply systems. Commands that simplified inventory tracking, visualized workflows, and eliminated unnecessary process steps saw measurable gains in readiness and efficiency. Sailors could quickly see where delays emerged and take action. Instead of managing complexity, they mastered it. Lean Six Sigma also reinforces a mindset of deliberate simplicity where each step in a process is questioned, refined, and made easier to teach and repeat. Instead of relying on expert intuition, it builds systems that anyone can execute with confidence. For naval leaders, simplifying warfare doctrine, briefs, and evolutions ensures every watchstander, regardless of rate or rank, knows their role and sees the bigger picture.

Take the Helm
- **Visualize the Plan** – Build tactical briefs around clear visuals, sequence, and top priorities. Train for execution, not just understanding.

- **Test for Transparency** – Ask your Sailors to explain key procedures in their own words. If they can't, the plan isn't ready.

Eyes on the Horizon
Simplify the plan and your team will deliver it with precision.

Day 295 – Turn Feedback Into Combat Power

The 7 Habits of Highly Effective People – Stephen R. Covey

"Seek first to understand, then to be understood."

Stephen R. Covey emphasizes that highly effective leaders first seek to understand before pushing their own point of view. This fifth principle of his challenges leaders to listen empathically, without filtering through ego or hierarchy. "Most people do not listen with the intent to understand. They listen with the intent to reply," he writes. Strategic leaders do the opposite. They mine every conversation for clarity, perspective, and insight. In naval operations, feedback from the deckplates often carries the clearest signal. That's why Admiral Michael M. Gilday, the 32nd Chief of Naval Operations, launched the "Get Real, Get Better" campaign. It was a fleet-wide initiative built on honest assessment and bottom-up feedback to drive operational learning. Commands were challenged to listen actively, identify root causes, and empower Sailors to own the process of improvement. Those that embraced this feedback loop saw measurable gains in maintenance, training, and morale, showing that Covey's model applies directly: when leaders listen deeply and without ego, they build trust. And when teams feel heard, they take greater ownership. Sailors fight harder for a command where their voice shapes the outcome.

Take the Helm
- **Act on What You Hear** – After high-tempo evolutions, hold structured feedback sessions at every level then act visibly on what's learned.

- **Institutionalize Listening** – Build feedback loops into drills, debriefs, and watch turnover to surface insight from every level.

Eyes on the Horizon
The next fight will be won by today's lessons. Listen well and turn feedback into advantage.

Day 296 – Know the Operational Environment

Sea Power – Admiral James Stavridis

"The sea is not a chessboard. It's a living, breathing organism shaped by politics, weather, alliances, and history."

Strategic leadership begins with situational awareness not just of your ship, but of the environment it operates in. Admiral James Stavridis underscores that naval operations unfold in complex theaters shaped by geopolitics, alliances, history, and diplomacy. Each maneuver does double duty: guiding steel and delivering a message. Leaders who understand this broader context make decisions that protect national interests and prevent miscalculation. Nowhere is that clearer than in the Bab al-Mandeb Strait, where U.S. Navy ships have operated near Yemen since 2015. On the surface, it's a narrow chokepoint. In reality, it's a flashpoint of tribal tensions, Iranian proxy activity, piracy threats, and commercial vulnerability. One aggressive maneuver or poorly timed hail could inflame regional tensions or signal unintended alignment. U.S. ships in the area must operate with tactical precision and cultural awareness, understanding not just the geography, but the dynamics shaping every contact. Every maneuver, message, and patrol must reflect that awareness. When you train your crew to understand the full battlespace, not just charts and contacts, but intent and consequence, you turn routine operations into strategic advantage. Because in today's Navy, understanding where you are is just as important as knowing why it matters.

Take the Helm
- **Teach Regional Context** – Include operational environment briefs in watch turnovers. Highlight regional actors, alliances, and geopolitical sensitivities.
- **Lead With Strategic Awareness** – Train watchstanders to consider how maneuvering, language, and reporting shape fleet perception not just contact management.

Eyes on the Horizon
You don't just sail in water. You sail in history, tension, and purpose. Know the environment and lead with intention.

Day 297 – Build a Warfighting Culture

With the Old Breed – Eugene B. Sledge

"War is brutish, inglorious, and a terrible waste. It takes simple men and demands of them absolute discipline, courage, and resilience."

Warfighting is about discipline lived daily. Eugene B. Sledge's account of fighting in the Pacific during World War II strips away any illusion of glory, showing instead the brutal requirement for steady readiness and resilience under fire. Sledge observed firsthand how the Marines who fought and survived on Peleliu and Okinawa endured because of the habits forged through rigorous training and unshakable purpose. Their weapons were cleaned obsessively. Their movements were rehearsed without complaint. In the worst conditions imaginable, discipline was survival. Even amid fatigue, fear, and chaos, Marines followed standard operating procedures, executed fire discipline, and held the line without constant direction. Their trust in each other was built not on words but on shared hardship and a common standard that didn't waver. Sledge recalled that the difference between life and death was often found in an attention to detail ranging from gear placement and trigger control to silent movement at night. That same mindset defines warfighting culture in the Navy. It's found in the watch teams who brief with urgency, the divisions that drill without prompting, and the commands that treat every evolution as preparation for the real fight. Culture isn't built in a message. A message starts the idea, but repetition cements culture in daily practice.

Take the Helm

- **Discipline Drives Readiness** – Reinforce daily habits that support warfighting. Bridge communication, CIC alertness, and system checks matter more than slogans.

- **Treat Readiness as Identity** – Use watch turnover, briefs, and drills to build pride in execution. Combat readiness starts with expectation.

Eyes on the Horizon

Culture is built in repetition. Make warfighting the norm and your team will meet any fight with quiet confidence.

Day 298 – Teach What You Want to Endure

Legacy – James Kerr

"The measure of a leader is what endures after they're gone."

James Kerr explores how the legendary New Zealand All Blacks built a culture that sustained excellence across generations. Their secret was transmission, not just training, but storytelling. Leaders like Richie McCaw and Colin Meads weren't remembered for stats alone, but for their humility, preparation, and service to the team with stories told and retold to reinforce identity. Kerr explains that when senior players swept the sheds, mentored juniors, and honored past greats in speeches and rituals, they weren't just building unity, but rather they were passing down values. The stories were preserved in locker room talks, pre-match huddles, and quiet one-on-one mentorship. This living narrative reminded each player they were more than an athlete but also stewards of a legacy. And their legacy was reinforced in language, rituals, and daily habits. These habits were small acts that embedded identity into every corner of their locker room. The same applies in the Navy. It's not the watch you stand today, but the standard you set that others maintain after you leave. Whether it's a qual program, a training rhythm, or a mindset of ownership, great leaders institutionalize excellence. What remains isn't just their guidance, but their influence woven into daily actions.

Take the Helm
- **Pass It On Intentionally** – When developing systems or processes, ask: Would this work if I transferred today? If not, build depth.

- **Lead With Legacy in Mind** – Mentor junior leaders to own and teach the standard. Legacy lives in how others carry your vision.

Eyes on the Horizon
Your greatest impact is what lasts after you're gone. Teach it right, build it strong and it will endure.

Day 299 – Strategic Leaders Look Out and Ahead

Sea Power – Admiral James Stavridis

"Look over the horizon. That's where the next challenge always comes from."

Strategic leadership requires daily execution as well as anticipation. Admiral James Stavridis urges leaders to lift their gaze and consider what's coming next, not just what's happening now. "The sea has always been a place of both opportunity and threat," he writes. "And it is the ability to look over the horizon that separates great maritime leaders from the merely competent." Leadership at sea, he argues, is not just about steering the ship but about shaping the waters ahead. Stavridis consistently ties naval leadership to global thinking: watching for geopolitical shifts, tracking emerging technology, and understanding how regional instability can impact even routine operations. As Supreme Allied Commander Europe, he regularly integrated naval posture with alliance coordination and diplomacy, using fleet presence to deter adversaries and reassure partners. He emphasizes that today's decisions, from port visits to patrol routes, are strategic signals. Embodying this mindset transforms crews from task-focused teams into forward-thinking warfighters. Strategic readiness is built by training Sailors to think, read, and speak in the language of the future fight.

Take the Helm
- **Build Forward Thinkers** – Task junior Sailors with briefings on future threats, tech trends, or evolving doctrine. Build curiosity into readiness.

- **Train for Tomorrow** – Push department heads to forecast not just next week, but next deployment. Lead the team into what's coming.

Eyes on the Horizon
Strategic leaders don't just react. Positive impact comes from their preparation, positioning, and purposeful guidance. Look ahead and bring your team with you.

Day 300 – The Fight Is Bigger Than You

The Mission, the Men, and Me – Pete Blaber

"It's not about your ego. It's about the mission. Always."

Pete Blaber lays out a clear philosophy shaped by years of special operations command: effective leaders subordinate ego to purpose. During the Battle of Fallujah in 2004, military leaders faced intense pressure to strike hard and fast. This desire to strike quickly was fueled by the brutal killing and public display of four American contractors, an act designed to provoke outrage and force a rash response. Political leaders and the public felt the pressure to respond decisively and visibly, to prove strength and reestablish control. But rushing in with overwhelming force risked turning the city's population against coalition forces, damaging international support, and fueling a broader insurgency. The ego-driven path was clear but wise leaders chose the harder road of restraint. Rather than pursue immediate destruction, they delayed operations to evacuate civilians and minimize collateral damage. That moment demonstrated Blaber's philosophy in action: when the mission calls for patience over pride, leaders must rise above ego. Coalition forces ultimately cleared the city with reduced civilian casualties, restoring control while maintaining international support, proving that tactical patience can achieve decisive outcomes without strategic fallout. When the demands of the mission rise, you rise with them, not for visibility or credit, but because it's what the team and the mission need.

Take the Helm
- **Subordinate Ego to Purpose** – Make decisions based on mission and team outcomes not personal gain or image. Let your Sailors see what selfless leadership looks like.

- **Set the Tone with Presence** – Be where you're needed most especially when it's inconvenient. Your actions define the culture.

Eyes on the Horizon
Real leadership stands on protecting what's right, serving with purpose, and building up those who follow. Put the mission first, and your legacy will follow.

CHAPTER 11:

LEADERSHIP BEYOND THE LIFELINES

Day 301 – Leadership Beyond the Lifelines

The Leader's Bookshelf – Admiral James Stavridis and R. Manning Ancell

"Leaders must read beyond their immediate field of vision, beyond the lifelines of their ship, into the vast ocean of ideas that shape the world."

The most enduring leaders operate on two frequencies at once: the here-and-now of daily operations and the long-range vision of institutional stewardship. As you rise in responsibility, your leadership must grow wider than your workspace. You're no longer just solving local problems. Now you're shaping systems and outcomes that extend well beyond your immediate command. "Leadership beyond the lifelines" means adopting a mindset that balances policy awareness with operational practicality and short-term needs with long-term vision. It means asking not just "What do we need to do?" but "What will this mean for those who come after us?" A powerful example came in 2019, when MCPON Russell Smith testified before Congress about the strain of extended deployments and the disconnect between operational demand and Sailor quality of life. His message wasn't reactive, but strategic, deliberate, and grounded in a broader understanding of what the Navy needed to retain trust and talent. He was reminding the institution that people are its most vital asset. His voice, shaped by decades of experience and deckplate credibility, carried beyond one command. It influenced how the Navy viewed readiness itself.

Take the Helm
- **Engage in Strategic Conversations** – Bring deckplate insight into discussions on policy, readiness, and Sailor welfare even when you're not in the room where decisions are made.

- **Mentor Across Commands** – Look beyond your division. Invest in peer mentorship, cross-platform relationships, and leadership development efforts that build institutional continuity.

Eyes on the Horizon
Real influence lives in how you protect, promote, and preserve standards and values across the Navy, not just within your workspace. True leaders carry the fleet forward by stewarding both its people and its purpose.

Day 302 – Teach Others to Teach

The Culture Code – Daniel Coyle

"Great cultures don't just tell people what to do. They build systems that teach them how to teach others."

The best leaders measure success by how capable their teams are when they step aside. Beyond the lifelines, this principle multiplies in power: the Navy thrives when knowledge is passed down with purpose, and Sailors are trained not just to qualify, but to coach. Warfare pins, qualification boards, and advancement exams all depend on this layered leadership structure. When senior leaders cultivate a culture where everyone is expected to teach the next tier, the system becomes resilient, scalable, and enduring even in the face of high turnover or operational strain. Daniel Coyle studied elite teams across disciplines and found that lasting excellence wasn't built on top-down instruction, but on peer-to-peer transfer of mastery. This is most visible in our qualification pipelines. A Sailor doesn't just earn a pin. They become responsible for preparing the next. But it's easy to let that culture fade when the pace picks up, during workups, INSURV prep, or deployment turnover, when the temptation is to get it done yourself rather than coach someone else through it. Leaders beyond the lifelines recognize this danger and intervene not by micromanaging but by reinforcing expectations, inspecting the standard, and making time to teach teaching. That's where real continuity lives.

Take the Helm
- **Build Teaching Into Qualification Standards** – Ensure your watch teams and qualifications emphasize mentorship responsibilities, not just task execution.

- **Inspect the Teaching Culture** – Ask not just who's qualified, but who's developing others and reward those who invest in future trainers.

Eyes on the Horizon
Leadership that shapes the next generation isn't judged by what you taught, but by how many new teachers you raised up. Strengthen the system by making development a duty.

Day 303 – Carry the Culture Forward

The War of Art – Steven Pressfield

"The amateur lives for the short term. The professional endures for the long haul."

Every command has a culture. Some are strong, others fragile. Some are cohesive and honorable while others are fractured and toxic. But the question isn't whether a culture exists, but whether leaders at every level take ownership of it. Steven Pressfield applies this principle through the lens of "Resistance," his term for the internal force that opposes creative and disciplined work. He contrasts the amateur, who caves to Resistance, whether in the form of procrastination, self-doubt, or distraction, with the professional, who expects Resistance and works anyway. He gives the example of the professional writer who sits down every day at the same time, regardless of inspiration, and does the work. The professional, he notes, "shows up every day, shows up no matter what, stays on the job all day," and "masters the technique." That's what command culture requires. Fragile and toxic cultures don't appear overnight. They take root when leaders, like amateurs, give in to Resistance. Each time we avoid a tough conversation, delay addressing a standards violation, or choose comfort over consistency, we surrender ground. Culture doesn't drift randomly. The direction follows whatever behavior leaders consistently allow. Leaders must steer it intentionally.

Take the Helm
- **Model the Standard Relentlessly** – Don't assume the culture will carry itself. Inspect, reinforce, and embody the tone you want your crew to reflect.

- **Name and Correct Drift** – When standards slip or core values are neglected, speak up. Silence is consent at senior levels.

Eyes on the Horizon
Each leader plays a role in keeping command culture strong or allowing it to erode. Be the kind of leader whose example keeps the culture strong.

Day 304 – Be the Fleet's Conscience

The Unforgiving Minute – Craig Mullaney

"Your integrity is the only thing that cannot be taken from you. It can only be surrendered."

At the enterprise level, your presence carries institutional weight. Your silence does too. As you rise in leadership, your role shifts from task manager to cultural bellwether. The higher you go, the more your words, and your omissions, shape policy and the next generation of leaders. In that sense, senior leaders become a "conscience" for the fleet. Craig Mullaney reflects on the weight of hard choices and when loyalty to subordinates, to values, and to the truth must all be balanced under pressure. At West Point, Mullaney faced a painful decision when a close friend violated the Cadet Honor Code by cheating. Bound by the code's "non-toleration" clause, he chose to report the incident, knowing it would lead to his friend's dismissal. The decision was agonizing, but it reinforced that integrity means upholding standards even when it hurts. The most respected leaders aren't those who flinch from hard conversations, but those who uphold integrity even when it isolates them. Whether it's reporting up the chain about personnel burnout or standing firm when morale programs are cut in favor of short-term metrics, your courage at the institutional level protects the present and preserves the Navy's long-term health.

Take the Helm

- **Stand the Moral Watch** – When decisions compromise core values, mission alignment, or Sailor welfare, be the voice that reminds leadership what truly matters.

- **Model Ethical Courage** – Encourage others to raise concerns early and create a climate where protecting the truth is a leadership expectation, not an exception.

Eyes on the Horizon
The Navy's conscience isn't found in a manual. How senior leaders carry themselves speaks louder about the fleet's conscience than any regulation ever could. Lead so that your voice strengthens trust, and your silence never casts doubt.

Day 305 – Connect the Deckplates to the Decisions

The Art of War – Sun Tzu

"Strategy without tactics is the slowest route to victory. Tactics without strategy is the noise before defeat."

At the strategic level, the greatest service you can provide is to connect decision-makers with operational realities. In one stark example, the Navy's push to improve resiliency and mental health support stemmed from the need to align policy with deckplate realities. Initial programs struggled until senior leaders began actively listening to junior Sailors about deployment fatigue, family stress, and the stigma around seeking help. Their feedback informed program design, which embedded mental health providers and implemented flexible counseling access, causing trust and effectiveness to improve. Sun Tzu's wisdom endures because it reminds us that vision and action alone are not enough. The fleet suffers when policies are made without understanding how they'll land on a watchbill or when new programs ignore the realities of maintenance cycles or family strain. As a senior leader, your stewardship means vetting good ideas before they become burdensome edicts. It means bringing up questions others may not know to ask: Will this policy improve trust? What tradeoffs will this change create? And most importantly, are we solving the right problem?

Take the Helm
- **Be the Translator Between Levels** – Use your position to help flag officers and policy makers understand how decisions will be received and implemented on the deckplates.

- **Protect Alignment** – Advocate for solutions that make sense strategically and tactically, even if it means slowing down to reassess.

Eyes on the Horizon
Leadership beyond the lifelines isn't about defending the status quo. Strong leaders focus on advancing both strategy and Sailors in step, never settling for the old ways. Connection is your greatest contribution.

Day 306 – Guard the Gate for the Next Generation

Man's Search for Meaning – Viktor E. Frankl

"For the world is in a bad state, but everything will become still worse unless each of us does his best."

As Sailors grow and organizations evolve, senior leaders become the gatekeepers of traditions and the expectations we pass on. It's your responsibility to guard what gets passed forward. In the concentration camps, Viktor E. Frankl recalls a powerful moment when fellow prisoners gathered quietly to observe a religious holiday, despite the grave risk. On one occasion, a group of men secretly lit a makeshift Hanukkah menorah fashioned from bits of scrap and stolen margarine, just to keep their spiritual traditions alive. In another scene, a prisoner gave up his bread ration to comfort someone else, an act of defiant humanity that preserved dignity and moral order in a place designed to erase both. These moments were about safeguarding meaning and values for those who might live to carry them forward. Frankl believed such acts proved that even in a world stripped of freedom, leaders could still choose how to respond with courage, compassion, and a deep sense of responsibility to others. Like Frankl's survivors, leaders who endure with purpose preserve the light for those who follow. It's staying committed to ethical promotions, challenging weak mentorship, and resisting the temptation to make decisions out of convenience. When each leader does their best, the institution stays worthy of those who come next.

Take the Helm

- **Preserve the Integrity of Advancement** – Don't rubber-stamp. Ensure those you advance reflect the Navy's values, not just a resume of accomplishments.

- **Set the Bar, Don't Lower It** – When standards are questioned, articulate why they exist and show the discipline to uphold them.

Eyes on the Horizon
Every Sailor you approve becomes part of your legacy. Guard the gate wisely. Your credibility is the first qualification they inherit.

Day 307 – Know the System, Then Shape It

The Lean Startup – Eric Ries

"A startup is not a smaller version of a big company. It's an institution that needs to be managed differently under conditions of extreme uncertainty."

The Navy is not a startup, but it can often feel like one. With evolving missions, shifting resources, and new systems fielded before full integration, as a senior leader, your power lies in knowing how the Navy works and how to make it work better. You must understand policies, resourcing pipelines, manpower systems, and readiness cycles deeply enough to identify gaps, inefficiencies, and opportunities for change. Eric Ries applies this principle by encouraging leaders to use validated learning, testing assumptions, gathering real-time feedback, and adapting quickly based on results. Innovation thrives when leaders understand their systems deeply enough to question them effectively and make informed pivots. This approach mirrors the role of senior naval leaders who must learn the intricacies of institutional processes in order to shape smarter, more resilient systems from within. You've seen what breaks, what works, and what never reaches the deckplates. When you provide feedback, share lessons learned, or support fleet initiatives, you offer more than anecdotes. You deliver strategic input that guides the Navy forward. But only if you use it.

Take the Helm
- **Master How Things Get Done** – Learn the process behind funding, staffing, and fleet-wide decisions so you can engage beyond complaints and offer real solutions.

- **Be a Reformer, Not Just a Resistor** – When a system fails Sailors, don't just work around it. Engage the people who can fix it and show them the evidence.

Eyes on the Horizon
You've earned the right to shape the system but that influence fades if you don't apply it. Know the gears of the machine, then improve how they turn.

Day 308 – Be the Anchor Across Commands

Leadership in War – Andrew Roberts

"It is character that counts… the quiet, steadfast kind that does not change from day to day or place to place."

As leaders move from command to command, they become a source of continuity in an ever-rotating Navy. Your ability to carry values and professionalism across platforms makes you an anchor. In a system defined by turnover, you become the bearer of culture. And when you embody integrity and steadiness, no matter the ship, the mission, or the CO, you build trust that transcends commands. During the Peninsular War, the Duke of Wellington upheld strict discipline among his troops, refusing to tolerate looting even when allied commanders were more lenient. His unwavering standards earned him deep respect from both his men and local populations across shifting fronts. Wellington's consistency became a stabilizing force in a chaotic, multinational campaign. As Andrew Roberts notes, it was this steadfast character, not just his tactical brilliance, that made him a trusted anchor across every command he led. This kind of unshakable leadership mirrors the role senior naval leaders play as anchors of institutional values across commands and deployments. Your presence should stabilize, not shift with the winds of new leadership. When you reinforce the Navy's core values uniformly, you become a trusted reference point, regardless of location or assignment.

Take the Helm

- **Carry the Culture with You** – Treat every new assignment as a chance to re-anchor Sailors in the Navy's core standards not just adapt to the climate you inherit.

- **Be a Consistent Leader in Every Setting** – Let your expectations, tone, and standards reflect the institution, not just the preferences of your current command.

Eyes on the Horizon
The most trusted leaders are rooted, not reactive. Be the steady force Sailors rely on when everything else is in motion.

Day 309 – Align Intent with Implementation

Principles – Ray Dalio

"The biggest mistake most organizations make is to confuse goals with the process for achieving them."

Misalignment between intention and execution is a common threat. A well-intentioned policy can fail miserably if the people enacting it don't understand its purpose or if it's applied inconsistently across commands. Leaders at the enterprise level must do more than support initiatives, they must translate them. That means connecting strategic goals to practical realities, anticipating confusion, and ensuring policies land on the deckplates with clarity, fairness, and accountability. As founder of the hedge fund Bridgewater Associates, Ray Dalio explains how he implemented real-time data tracking, open dialogue, and iterative testing to ensure that decisions made in leadership meetings translated into the intended outcomes at the operational level. Bridgewater grew to become the largest hedge fund in the world, and managed over $150 billion at its peak, with strong performance through both bull and bear markets. This relentless focus on aligning actions with intent closely parallels the Navy's need to ensure that command policies are clearly understood and consistently applied across all levels. As a senior leader, your responsibility is to reduce that drift. When Sailors understand why a change is happening and how it supports the larger mission, they buy in. When leaders adjust execution with clarity and humility, the institution becomes stronger.

Take the Helm
- **Translate Vision into Action** – When new guidance rolls out, explain the purpose to your Sailors and help them see how it supports broader Navy goals.
- **Close the Loop** – Gather honest feedback from the deckplates on new programs or policies and pass it up the chain to refine implementation.

Eyes on the Horizon
Policies don't build trust. People do. Senior leaders turn vision into reality by making sure the execution honors the intent.

Day 310 – Invest in What Outlasts You

The War of Art – Steven Pressfield

"The professional arms himself with patience, not only to give the stars time to align in his career but to keep himself from flaming out."

Institutional leadership means thinking beyond personal recognition or immediate results. It's about building structures and processes that will support generations of Sailors long after you've transferred or retired. Senior leaders who steward legacy systems, training pipelines, qualification boards, and policy reviews understand that real change is often slow and thankless. But they stay engaged because they know the long-term health of the Navy depends on consistent, principled investment, and not short bursts of energy tied to a FITREP cycle. Steven Pressfield draws a sharp contrast between amateurs, who chase quick wins and external validation, and professionals, who commit to a long, often unseen path of mastery and contribution. He emphasizes that true professionals show up every day, not for applause, but out of devotion to the craft and a belief in its long-term value. This mindset is especially relevant to institutional leaders who must invest in systems and reforms that may not yield results during their tenure. The leaders who resist the temptation to "just get through the tour" and instead work to institutionalize progress are the ones whose impact is still felt long after the plaques come down.

Take the Helm

- **Champion Systemic Improvements** – Volunteer for working groups, feedback reviews, and policy rewrites that will benefit the fleet even if they don't benefit your resume.

- **Endure with Purpose** – Recognize that meaningful institutional progress takes time. Your role is to plant seeds others will water.

Eyes on the Horizon

A tour ends. A system remains. If you want your influence to outlive your assignment, build for the Sailors you may never meet.

Day 311 – Elevate Professional Standards at Scale

The Five Dysfunctions of a Team – Patrick Lencioni

"Great teams do not hold back with one another. They are unafraid to air their dirty laundry."

Institutional health depends on shared professional standards, enforced consistently across commands. When units or leaders operate with wildly different expectations, credibility erodes. The job of senior leaders is not only to meet high standards, but to elevate them fleetwide. That means modeling integrity in evaluations, confronting poor performance without delay, and holding peers accountable in professional forums. Patrick Lencioni emphasizes that fear of conflict, especially among senior leaders, is a key reason teams fail to uphold high standards. He explains that when leaders avoid holding their peers accountable to maintain superficial harmony, they unintentionally endorse mediocrity and allow performance to decline. This principle is reflected in the Navy's decision to publicly hold senior leaders accountable for misconduct or toxic command climates, often resulting in reliefs for cause or formal administrative action. Such high-profile examples remind us that standards don't stop at rank and that silence from senior leaders amounts to complicity, not neutrality. When leaders confront uncomfortable truths and act decisively, they protect the institution and raise the bar across the fleet. If you want the Navy to function as a high-performing system, you must be willing to elevate the conversation.

Take the Helm
- **Don't Normalize Mediocrity** – When a program, leader, or division is visibly underperforming, raise the issue through the appropriate channel even if it's uncomfortable.

- **Set the Tone in Professional Spaces** – Speak plainly, hold your ground, and make high standards the norm, not the exception, in working groups and leadership councils.

Eyes on the Horizon
The institution rises or falls on what leaders choose to tolerate. Raise the standard and raise others with it.

Day 312 – Think in Systems, Act with Urgency

Principles – Ray Dalio

"To be effective, you must not let your need to be right be more important than your need to find out what's true."

At the institutional level, effective leaders recognize that problems rarely exist in isolation. A missed deadline might reveal a training shortfall. A toxic climate could signal systemic neglect. When you think like a systems steward, you stop chasing symptoms and start identifying root causes. But systems thinking must be paired with urgency. If you only analyze and never act, good intentions become stagnation. Senior leaders must scan wide, dive deep, and then move decisively with data-informed solutions that impact not just a command, but the enterprise. The Navy's use of readiness systems like Training and Operational Readiness Information Services (TORIS) and Joint Unit Planning and Integration for Training Readiness (JUPITER) exemplifies this systems-thinking approach in action. TORIS revealed key shortcomings in how training readiness was tracked, highlighting issues such as data silos, inconsistent reporting, and limited enterprise-level visibility. In response, JUPITER was developed as a comprehensive solution that integrates real-time data across platforms, enabling leaders to identify root causes and implement meaningful, timely corrections. This aligns with Dalio's principle that true effectiveness comes from connecting insights across the system and acting decisively. Your value as a leader lies in not only recognizing patterns and linking causes to outcomes but taking action to drive enterprise-level improvement.

Take the Helm

- **Apply Root-Cause Thinking** – Don't just fix symptoms. Ask *why* five times to trace problems to their source and recommend sustainable fixes.

- **Move from Insight to Action** – Build the habit of following analysis with implementation. Don't let good ideas die in the discussion phase.

Eyes on the Horizon
Systems survive on insight but thrive on execution. See broadly, act boldly, and don't confuse thinking with leading.

Day 313 – Represent the Institution Well

Fearless – Eric Blehm

"Reputation, like honor, cannot be given. It must be earned daily."

When you operate at the institutional level, you carry the weight of the Navy's reputation. Enterprise-level leaders are ambassadors of the Navy's values, mission, and professionalism. The tone you set, the stories you tell, and the standard you model all become part of how the Navy is understood by those inside and outside the lifelines. Master Chief Carl Brashear exemplified this principle through his unwavering commitment to the Navy's values in the face of extraordinary adversity. As the first amputee to be recertified as a Navy diver in 1968, and later the first African American Master Diver in 1970, Brashear overcame systemic racism, career-threatening injury, and institutional doubt, not by demanding recognition, but by embodying professionalism and relentless perseverance. His conduct, both in and out of uniform, reflected the highest standards of the service and inspired generations of Sailors. Brashear's life was a living testament that the Navy's reputation is upheld not by slogans, but by the character of those who wear the uniform. His legacy continues to define what it means to be a true ambassador of the institution. Carrying the Navy's reputation means leading with integrity in every moment, because your presence speaks for something far greater than yourself.

Take the Helm
- **Be a Model of Navy Standards** – Speak, write, and lead in a way that reflects professionalism, clarity, and calm, especially in high-visibility settings.

- **Engage With Strategic Awareness** – Treat every external-facing interaction as a chance to strengthen the Navy's image and Sailor trust.

Eyes on the Horizon
You don't just wear the uniform. Every action reflects the values of the institution behind it. Carry it with the consistency and class that our Sailors and nation deserve.

Day 314 – Create Space for Innovation to Breathe

The Lean Startup – Eric Ries

"If we're building something that nobody wants, it doesn't matter if we're doing it on time and on budget."

One of the most overlooked responsibilities of senior leadership is protecting innovation from being crushed by bureaucracy. Are Sailors encouraged to challenge the process? Are good ideas surfaced and tested or buried by "we've always done it this way?" At IMVU, an early social networking startup where Eric Ries served as Chief Technology Officer, he implemented the "build-measure-learn" feedback loop, where small, fast tests replaced drawn-out development cycles ensuring ideas were validated by real user needs before being scaled. His experience reinforces that leaders must protect creative initiatives from bureaucratic drag and allow teams to test, adapt, and iterate. Institutional leaders must balance process discipline with adaptability. That means championing junior-led improvement projects and being willing to push change even when it disrupts entrenched hierarchies or legacy systems. The political discomfort often comes from challenging senior stakeholders or reallocating resources from legacy programs to unproven ideas. It means risking criticism for deviating from precedent or defending a junior Sailor's initiative in front of skeptical peers. Innovation takes root when leaders stay engaged long enough to see rough ideas refined into real solutions. If you want Sailors to think boldly, show them their ideas will be heard, tested, and supported from the top.

Take the Helm

- **Reduce Bureaucratic Drag** – Actively seek out approval bottlenecks or outdated SOPs that stall good ideas then advocate for process improvement.

- **Sponsor Experiments, Not Just Metrics** – Give Sailors time and room to test improvements, even if the outcome isn't perfect on the first try.

Eyes on the Horizon

If nothing can change under your watch, you're not leading. Stasis sets in when initiative is replaced by routine. Innovation requires space, and leaders must protect that space intentionally.

Day 315 – Steward the Profession, Not Just the Platform

Leading with the Heart – Mike Krzyzewski

"You don't just coach a team. You coach a program. You coach a legacy."

At higher levels of leadership, your focus must expand beyond immediate wins or platform performance. Great enterprise-level leaders think institutionally. They ask what kind of Navy they're reinforcing, not just whether their ship is performing well. Coach K recounts how he gave Grant Hill significant leadership responsibility early in his basketball career, not because of talent, but because of his poise, maturity, and decision-making. He involved Hill in key team conversations and trusted him to model behaviors for upperclassmen, showing that leadership was about character, not seniority. By trusting Hill with leadership responsibilities as a freshman, Coach K accelerated his development into a poised, unselfish, and team-centered player. Hill's maturity and calm under pressure became a stabilizing force for the team, contributing directly to Duke's national championships in 1991 and 1992. This mirrors the role of senior Navy leaders who must think beyond platform readiness and instead focus on shaping the enduring character and competence of the institution itself. You're shaping how we grow Chiefs, prepare officers, mentor enlisted talent, and enforce the values that define our identity. The true measure of your leadership is what you leave behind in the people and culture you shape.

Take the Helm
- **Uphold Community Standards** – Participate in boards, panels, and rating discussions that influence how leaders across the fleet are selected and evaluated.

- **Mentor With the Profession in Mind** – Develop others not just to succeed in their current roles, but to carry the Navy's core values into their next ones.

Eyes on the Horizon
Leaders serve more than commands. Upholding the profession means modeling excellence beyond immediate results. Your conduct should elevate the craft of leadership for everyone who follows in uniform.

Day 316 – Enforce Alignment Across Echelons

The Dichotomy of Leadership – Jocko Willink and Leif Babin

"If your team doesn't understand the mission, you haven't led."

It's not enough for your command to understand the mission. The commands around you must understand it too. Misalignment creates redundancy, confusion, wasted effort and worse, it creates contradictory guidance that erodes trust. Admiral Ernest King's leadership during World War II exemplified enforcing alignment across echelons by unifying disparate naval operations under a cohesive global strategy. As both Chief of Naval Operations and Commander in Chief, U.S. Fleet, he centralized authority to ensure that actions in the Atlantic, Pacific, and Mediterranean were synchronized with overall national objectives. King's ability to align policy, operations, and execution across commands helped turn a fractured wartime posture into a unified and effective naval campaign. He understood that even small misalignments between the Atlantic and Pacific fleets could have global consequences. Senior leaders today must do the same: regularly assess whether coordination between adjacent commands, TYCOMs, or warfare communities truly supports unified intent or reflects siloed priorities. When you detect misalignment between policy and practice, between what's written and what's lived, you have a duty to reconcile it. Sometimes that means asking hard questions. Other times it means reinforcing shared understanding. Always, it means ensuring that Sailors aren't caught in the friction between competing agendas.

Take the Helm

- **Audit for Alignment** – Regularly assess whether your team's efforts and neighboring commands' efforts actually support shared strategic goals and not just local milestones.

- **Clarify the Purpose Across Boundaries** – Ensure that operational priorities and guidance are communicated in a way that aligns horizontally across platforms, not just vertically up and down the chain.

Eyes on the Horizon

Senior leaders align purpose. When commands pull in the same direction, the Navy moves with strength and speed.

Day 317 – Strengthen Institutional Memory

Sea Power – Admiral James Stavridis

"The sea has a long memory, and so must those who lead upon it."

One of the greatest risks in any large organization is forgetting what it has learned the hard way. Leaders beyond the lifelines become custodians of institutional memory, those who remember the context behind a policy, the unintended consequences of a past change, or the reason why a standard exists. What some call nostalgia is really an informed respect for history, honed by wisdom. When senior leaders pass on not just the *what* but the *why*, they help the Navy avoid repeating mistakes and preserve lessons that may not be captured in any instruction. Admiral Stavridis highlights Admiral Nelson's legacy at the Battle of Trafalgar as a powerful example of institutional memory in action. Nelson's emphasis on decentralized command, clear intent, and bold initiative became foundational principles passed down through generations of naval leaders. Stavridis explains that remembering and studying such pivotal moments preserves not just tactics, but the deeper leadership values that shaped them. This reinforces the idea that naval leaders must be stewards of historical insight, ensuring that critical lessons remain alive in doctrine, training, and decision-making. If we want a better future, we must intentionally preserve the wisdom that shaped our present.

Take the Helm

- **Preserve Context in Policy** – When writing or updating guidance, include background on why decisions were made, not just what actions are required.

- **Pass Down Lessons, Not Just Instructions** – Share the reasoning, historical examples, and past failures that shaped your current systems so others lead with insight.

Eyes on the Horizon
Leaders carry experience as well as context. Preserve the Navy's memory so others can navigate the future with clarity.

Day 318 – Lead Through Institutional Transitions

Emotional Intelligence – Daniel Goleman

"In a very real sense we have two minds, one that thinks and one that feels."

Institutional transitions such as new programs, reorganizations, or policy shifts, test a leader's ability to maintain trust while navigating uncertainty. When change rolls downhill, Sailors look to senior leaders not just for orders, but for emotional grounding. Emotional intelligence is essential here. The decommissioning of *USS Enterprise* in 2012 was a deeply symbolic transition, marking the retirement of the Navy's first nuclear-powered aircraft carrier after more than 50 years of service. Leaders recognized that this was not just a procedural event but also an emotional one, which affected thousands of Sailors who had served aboard and viewed the ship as a defining part of their identity. By organizing formal ceremonies, inviting past crew members, and preserving the ship's legacy through detailed storytelling and media outreach, leadership ensured the transition honored both history and sentiment. The decision to pass the *USS Enterprise* name to a future Gerald R. Ford–class carrier also helped maintain continuity and institutional pride. These actions reflected a high degree of emotional intelligence, acknowledging grief, celebrating legacy, and reinforcing the enduring values the ship represented. It demonstrated that successful institutional transitions require leaders who can guide people through change with empathy, clarity, and reverence. When handled well, transitions become opportunities to reinforce professionalism and unity.

Take the Helm

- **Anticipate the Fallout** – When major institutional changes are announced, assess how they'll practically affect your team and build the support plan before the impact hits.

- **Speak With Unified Clarity** – Even if you disagree with the change, present the information with professionalism and context. Conflicting messages create chaos.

Eyes on the Horizon
In times of change, clarity and steadiness from senior leaders turn disruption into momentum.

Day 319 – Defend the Integrity of Recognition

The 12 Rules for Life – Jordan B. Peterson

"You can't have the confidence to do difficult things unless you speak the truth."

Recognition programs, awards, and qualifications carry deep institutional weight when used correctly. Senior leaders must defend the integrity of recognition at scale. That means standing firm when pressure mounts to approve weak packages. It means challenging inflated write-ups, incomplete qualifications, or end-of-tour awards that don't match the impact. The Navy and Marine Corps Achievement Medal (NAM) is intended to recognize noteworthy performance, not routine duty or time-in-service. When NAMs are awarded automatically at the end of a tour or for expected responsibilities, it dilutes their meaning and undermines those who earned the award through truly exceptional impact. Defending the integrity of the NAM requires leaders to apply consistent standards, resist pressure to "pad" packages, and ensure the write-up reflects measurable contributions. As Jordan B. Peterson emphasizes, recognition must be grounded in truth. When it's not, it erodes morale, breeds resentment, and weakens the institution's credibility. When leaders approve recognition that isn't earned, they dilute the message to those who did earn it. Defending the integrity of recognition protects morale in the long term even when it causes friction in the short term.

Take the Helm

- **Honor the Standard, Not the Pressure** – Don't approve awards or quals out of convenience. Ensure the recognition reflects genuine merit and alignment with Navy values.

- **Speak Up When It's Watered Down** – If you see a trend of recognition being used as appeasement or inflation, bring it to the forefront with command leadership.

Eyes on the Horizon

When everything is special, nothing is sacred. Defend the value of recognition so that Sailors can trust it means what it says.

Day 320 – Clarify Roles, Prevent Drift

Team of Teams – General Stanley McChrystal

"In complex environments, resilience often matters more than efficiency."

In large commands or across multi-unit structures, confusion about roles and responsibilities can be just as damaging as incompetence. Drift happens when teams act without clarity, overlap in authority, or assume someone else is accountable. Senior leaders must be especially vigilant in clarifying roles, not just within their direct reports, but across departments, detachments, or commands. This is particularly important during inspections, high-tempo operations, or institutional reviews. General Stanley McChrystal recounts how early Joint Special Operations Command operations in Iraq were hindered by role confusion between military units, intelligence agencies, and coalition partners, leading to stalled missions and missed opportunities. Proposing a daily Operations and Intelligence briefing across such a wide and bureaucratic network wasn't initially welcomed. Many saw it as inefficient, overly ambitious, or unnecessary. But McChrystal's resilience in pushing through that resistance was key. He knew that shared awareness and clearly defined responsibilities would outweigh short-term friction. That shift ultimately created a more adaptive and aligned force, capable of responding in real time because every team knew their role and how it fit into the mission. Resilience in leadership is about challenging dysfunction, pushing for systems that work, and holding the line when clarity is unpopular. A simple reminder that in fast-moving environments, clarity is what turns good intentions into coordinated action.

Take the Helm

- **Define Boundaries and Interfaces** – Ensure each team knows not only what they own but where they intersect with others, especially during mission-critical phases.

- **Check for Silent Gaps** – Look for essential tasks that may have fallen into the cracks between roles and assign clear ownership.

Eyes on the Horizon

Confusion breeds failure faster than inexperience. Lead by bringing clarity where complexity threatens cohesion.

Day 321 – Be a Builder of Bench Strength

Radical Candor – Kim Scott

"The essence of leadership is not getting overwhelmed by responsibility, but empowering others to share it."

Enterprise-level leadership means you're no longer the only answer. As the builder of answers in others, you're creating a deep bench of capable Sailors who can lead with confidence. That takes intentional development. Kim Scott recounts her time at Google, where she learned the importance of combining direct feedback with genuine personal care, most notably under the mentorship of Sheryl Sandberg. Sandberg once gave Scott radically candid feedback that her communication was coming off as dismissive, not to criticize her, but to help her grow into a more effective leader. Sandberg offered tangible support by connecting Scott with a speaking coach, reinforcing that real leadership means investing in others' long-term growth. This moment became foundational for Scott's concept of *Radical Candor*, the ability to care personally while challenging directly. The lesson applies directly to bench-building: leaders grow other leaders not by avoiding hard truths, but by delivering them with respect, purpose, and follow-through. Great leaders invest in people not by shielding them, but by preparing them. This looks like mentoring LPOs into Department LCPOs, grooming E-5s to become Chief-select ready, and preparing Sailors to lead watch teams without direct oversight. When you build leadership capacity around you, you reduce operational risk and increase long-term resilience. Bench strength is your legacy in motion.

Take the Helm
- **Grow Your Replacement Early** – Identify promising leaders at every level and assign them increasing responsibility with mentorship, not micromanagement.

- **Coach With Candor** – Give your future leaders the honest feedback they need to develop, not just the encouragement they want to hear.

Eyes on the Horizon
Multiply your influence by strengthening the leadership bench around you.

Day 322 – Understand the Levers of Readiness

Sea Power – Admiral James Stavridis

"You can't surge trust. You can't surge leadership. And you certainly can't surge seamanship."

Naval readiness is an ecosystem. Manning, maintenance, training, funding, and morale all act as interdependent levers. When one fails, the rest are strained. This means asking questions about long-term sustainability, trade-offs between urgent and important, and what gaps are being quietly accepted because they're too complex to fix. Admiral James Stavridis reflects on his command of *USS Barry* during a high-tempo Mediterranean deployment. Despite the ship receiving accolades and external praise, Stavridis highlights that this outward success masked the invisible, deliberate preparation behind it. Months of integrated training cycles, rigorous engineering and combat systems upkeep, and a clear command climate rooted in trust were what allowed the crew to perform under pressure. While *USS Barry* performed well by external metrics, Stavridis emphasizes that this success was only possible because of sustained investment in crew training, preventative maintenance, and cohesive leadership, elements that aren't easily measured but are essential to true readiness. He notes that without trust within the team and deep professional competence, the ship's performance could have quickly faltered under operational pressure. This example illustrates his point that readiness cannot be surged and must be deliberately built and preserved over time. Institutional stewards must be the ones who speak up when the readiness picture looks good on paper but brittle in reality.

Take the Helm
- **Know What Drives Readiness** – Develop fluency in the Navy's readiness model. Understand how manning levels, training cycles, and maintenance backlogs interact.

- **Surface Quiet Risks** – Raise visibility on readiness trade-offs, or "accepted risks," that may have become normalized but threaten mission success.

Eyes on the Horizon
Readiness is a balance to maintain, not a box to check. Lead by seeing the whole system, not just the scoreboard.

Day 323 – Prioritize What the Institution Needs

Resilience – Eric Greitens

"What you do when you don't have to will determine what you will be when you can't help it."

Priorities are often shaped by operational tempo and immediate tasking. Leaders must think differently and anticipate what's needed. Sometimes, that means championing unpopular initiatives. Other times, it means pushing back on misallocated energy and programs that sound good in theory but distract from core readiness. In the early 2000s, the Navy replaced the traditional SWOS Division Officer schoolhouse training with a shipboard, self-guided program officially known as SWOS-at-Sea, which was informally criticized across the fleet as "SWOS-in-a-box." Junior officers were expected to complete training via CD-ROMs aboard their ships while simultaneously qualifying for watches and leading divisions. Although intended to save time and resources, this approach quickly revealed significant shortcomings. Fleet leaders across the waterfront documented gaps in junior officer performance and spoke candidly in formal reports, informal feedback loops, and community forums. Some met resistance from advocates of the new model, who emphasized cost savings and modern delivery methods. But experienced leaders pushed back, insisting that warfighting standards, not budget or convenience, must drive officer development. Their advocacy eventually led to the reinstatement of in-residence SWOS training, restoring structured instruction in navigation, seamanship, leadership, and systems. It was a hard correction, but the right one, proving that long-term resilience often starts with principled friction.

Take the Helm
- **Champion What Matters Most** – Focus energy on enduring programs and improvements that advance warfighting readiness and Sailor well-being, not vanity metrics.

- **Decline the Distracting** – Say no to efforts that dilute strategic focus, even if they're politically convenient or temporarily popular.

Eyes on the Horizon
When you lead for the institution, not just your inbox, you serve the mission at scale. Purposeful prioritization is your most powerful act of stewardship.

Day 324 – Influence Morale Without a Microphone

Emotional Intelligence – Daniel Goleman

"The leader's mood is quite literally contagious, spreading quickly and inexorably throughout the business."

Your influence on morale comes from how you carry yourself during setbacks, how you respond to bad news, how you handle silence in meetings, and how you walk the passageways after a long day. Senior leaders often underestimate the emotional tone they set, not with words, but with behavior. Your emotional regulation becomes your Sailors' climate stabilizer. Daniel Goleman recounts the story of a hospital executive whose quiet presence had a profound effect on organizational morale through his nonverbal cues and emotional tone. In one instance, the executive entered a trauma planning session visibly agitated after reading a dire patient prognosis. His tension set off a chain reaction: senior clinicians grew short with one another, nurses hesitated to speak up, and the team missed key coordination points for a high-risk procedure. The executive later reflected that his demeanor had disrupted the room before a word was even spoken. In contrast, during a subsequent incident involving multiple critical casualties arriving from an accident, he remained composed and engaged. His steady presence helped the team stay focused, communicate clearly, and execute with cohesion. This directly applies to senior Navy leaders, whose visible composure, or lack of it, can either stabilize or destabilize morale across commands without ever needing a microphone.

Take the Helm

- **Project Stability, Not Showmanship** – Maintain a calm, focused presence even when tensions rise. Your body language, tone, and timing influence morale more than motivational quotes.

- **Support Below Without Undermining Above** – If Sailors are struggling with higher-level decisions, acknowledge their stress without eroding confidence in the institution.

Eyes on the Horizon

When you can no longer fix every problem directly, you lead through tone. Make yours a steadying force that others can trust.

Day 325 – Reinforce the Purpose at Scale

The Art of War – Sun Tzu

"When the leader is morally strong and disciplined, and the troops are treated with fairness, they will be loyal and willing to die for him."

As leaders move higher in the institution, they gain distance from daily execution. But that doesn't lessen their responsibility. If anything, it increases their obligation to reinforce the purpose behind the Navy's most demanding tasks. Leaders beyond the lifelines must translate the strategic vision into moral clarity that reaches the deckplates. The Navy's Advancement Modernization initiative exemplifies the principle of reinforcing the purpose at scale by shifting from a test-centered system to one that values sustained performance, leadership, and mission readiness. While the changes created uncertainty among Sailors accustomed to legacy advancement models, commands saw stronger buy-in when they clearly explained the rationale of improving fairness, recognizing consistent performers, and aligning advancement with fleet needs. By connecting policy changes to long-term institutional goals, senior leaders helped Sailors understand that the purpose wasn't to complicate advancement, but to make it more reflective of real-world contributions. This effort reinforced that trust and transparency matter as much as the policy itself. In times of change, Sailors need to hear the purpose not just from the top but echoed clearly through every level of leadership.

Take the Helm

- **Connect Strategy to Sacrifice** – When Sailors express frustration with policies or changes, explain the operational or institutional reasoning behind them in plain, mission-focused terms.

- **Repeat the Message with Purpose** – Revisit messages regularly especially during high-tempo periods, new rollouts, or extended hardship.

Eyes on the Horizon
You can't ask Sailors to endure without offering meaning. When you clarify the purpose, you carry them farther than directives ever will.

Day 326 – Set the Standard Among Peers

Radical Candor – Kim Scott

"The fastest path to building trust is to care personally and challenge directly."

The institution relies on senior leaders to hold each other accountable, even when it's uncomfortable. If you let a fellow department head dismiss a Sailor's concern or excuse a poor decision, you don't just condone it, you also normalize it. Leadership beyond the lifelines means setting the tone in the spaces where silence is often the easiest choice. Kim Scott shares how she failed to confront a peer at Google whose toxic behavior, quietly eroded team morale. She recounts how he routinely belittled teammates in meetings, dismissed ideas without listening, and created an atmosphere of fear and insecurity. Despite his strong performance metrics, his behavior undermined collaboration and stifled innovation, violating the very values the team was meant to uphold. Scott admits that her failure to challenge this conduct stemmed from a reluctance to "rock the boat," even though it conflicted with her responsibility to foster a healthy culture. She reflects on this as a failure of her leadership, not because she lacked authority, but because she chose silence over accountability among peers. The experience taught her that setting the standard means having the courage to challenge colleagues when their actions conflict with the team's values. It's a reminder that at senior levels, leadership is measured as much by how we influence our peers as by how we manage those below us.

Take the Helm

- **Speak Truth Peer-to-Peer** – Challenge your peers respectfully when standards, decisions, or behavior conflict with Navy values or mission alignment.

- **Protect Below by Leading Beside** – Don't force junior Sailors to call out senior failures. Correct them at your level, where it belongs.

Eyes on the Horizon

Leaders who challenge peers strengthen the institution. If you tolerate it in your circle, you authorize it everywhere else.

Day 327 – Use Authority with Discipline

Leadership and Self-Deception – The Arbinger Institute

"When I see others as people, I have a different sense of what I should do."

The higher you rise in the Navy, the more authority you carry and the more dangerous it becomes if unchecked. Senior leaders are granted enormous influence: over policy, people, evaluations, and decisions that ripple across commands. The authority to enforce must be balanced by empathy, context, and principle because power that isn't grounded in perspective becomes tyranny dressed in uniform. Following the attack on Pearl Harbor, Admiral Chester Nimitz assumed command of the battered Pacific Fleet at a time when morale was low and pressure for swift retribution was immense. Despite holding near-total authority and facing political calls for immediate accountability, Nimitz chose not to scapegoat or publicly relieve senior officers. Instead, he exercised measured judgment, recognizing the need to preserve experience and restore confidence rather than impose fear. By demonstrating restraint and empathy, he reinforced trust in leadership and focused the fleet on recovery and strategic clarity. His disciplined use of authority stabilized the Navy during one of its darkest hours and laid the foundation for decisive victories at Coral Sea, Midway, and beyond. Nimitz's leadership exemplifies that the most powerful leaders are those who wield authority with moral clarity, fairness, and self-control. Discipline in judgment ensures that your power strengthens the organization instead of eroding its trust.

Take the Helm
- **Use Your Rank to Serve, Not Prove** – Let your decisions reflect measured professionalism, not the unchecked impulse to be right or stay in control.
- **Pause for Perspective** – Before exercising judgment, ask who will be affected and whether you've considered all sides fairly.

Eyes on the Horizon
At the institutional level, how you use authority defines the culture. Restraint, empathy, and fairness are the marks of trusted leadership.

Day 328 – Protect Institutional Credibility

The Culture Code – Daniel Coyle

"Trust is not built by grand gestures, but by small, consistent signals."

In a large institution like the Navy, trust takes years to build and moments to lose. Senior leaders hold the keys to that trust. Whether you're influencing policy, approving communication, or setting the tone in forums like leadership conferences or command boards, your decisions either reinforce or diminish institutional credibility. Credibility is about owning mistakes, staying consistent, and refusing to tolerate shortcuts in character. Sailors will forgive a leader who tries and errs. They won't forget one who deflects or deceives. In 2016, the Navy announced the elimination of traditional enlisted ratings as part of a modernization effort intended to promote flexibility and align with civilian career systems. However, the move was met with overwhelming backlash from Sailors who felt the change disregarded heritage, identity, and community cohesion, all of which were key components of institutional trust. Rather than doubling down, Navy leadership, including then-CNO Admiral John Richardson, listened to the fleet and reversed the decision, restoring ratings in early 2017. This reversal was not a sign of weakness. Choosing to acknowledge missteps, value feedback, and place credibility above pride showed true discipline and leadership. By choosing transparency and responsiveness, the Navy preserved its institutional legitimacy and demonstrated that leadership remains accountable to the force it serves.

Take the Helm

- **Be Predictable in Principle** – Align your words and actions, especially when it's unpopular or inconvenient. Credibility requires consistency.

- **Enforce Transparency from the Top** – Ensure selection boards, command messaging, and decisions that affect careers are handled with fairness and visible integrity.

Eyes on the Horizon

Credibility is your currency at the institutional level. Earn it through steadiness. Protect it through principle.

Day 329 – Sustain Trust Through Tough Decisions

The Unforgiving Minute – Craig Mullaney

"Leadership is the art of making people want to do what must be done."

When you operate at the strategic level, every decision has weight, many of which require having winners and losers. Assignments, funding, billets, and program cuts all force trade-offs. The measure of your leadership isn't just what you decide, but how you handle the impact. In 1799, during the Quasi-War with France, Captain Thomas Truxtun commanded *USS Constellation* in a bold and calculated engagement against the French frigate *L'Insurgente*. He won a clear tactical victory, the U.S. Navy's first major triumph at sea. But afterward, Truxtun faced a different kind of decision: whether to treat captured French sailors with dignity or exact vengeance in the name of morale and retribution. Some among his crew, worn down by tension and eager for retribution, expected harsh treatment. Truxtun chose instead to uphold humane standards, risking momentary discontent to reinforce the Navy's emerging ethical foundation. He lost short-term favor with a few but preserved long-term trust by making the harder right over the easier wrong. In later decisions, he would bypass potential glory to protect his crew and the mission, demonstrating that real leadership isn't about popularity or posturing, but about values, consistency, and transparency. Trust must be earned, protected, and reaffirmed, especially when decisions create visible winners and losers. When your Sailors see you sacrifice personal gain for principle, that's when trust becomes unshakable.

Take the Helm
- **Be Transparent in Trade-Offs** – When making tough calls, explain your reasoning clearly to those affected. Don't hide behind rank or process.

- **Carry the Decision With You** – Own outcomes publicly even when the reaction is mixed. Don't shift blame or deflect consequences downward.

Eyes on the Horizon
Hard choices define senior leadership. Make yours with integrity, and you'll carry trust longer than consensus ever could.

Day 330 – Steward the Institution Quietly

The War of Art – Steven Pressfield

"The professional keeps his eye on the doughnut and not on the hole. He focuses on the work and allows rewards to come, if they will."

As a senior leader, your charge is to preserve the trust, traditions, and systems that allow the Navy to function under stress. Rear Admiral Wayne E. Meyer, known as the "Father of Aegis," the Navy's premier integrated air and missile defense system, exemplified the principle of stewarding the institution quietly through his decades-long leadership of the Navy's most transformational combat system development. Rather than seeking personal recognition, Meyer focused relentlessly on building a system that would work, employing his "Build a Little, Test a Little, Learn a Lot" philosophy to ensure Aegis was robust, scalable, and integrated into every layer of the fleet. He navigated complex inter-service politics, contractor relationships, and shifting defense priorities with quiet discipline, always keeping the mission, fleet survivability and superiority in the missile age at the center. Meyer refused shortcuts, often delaying milestones or resisting premature public praise to protect the program's integrity and long-term viability. His leadership created not just a system, but a culture of technical rigor and institutional stewardship that still defines surface warfare development. Meyer's enduring legacy is not in visibility, but in the generations of Aegis-equipped warships quietly defending the fleet, each one a testament to leadership focused on outcomes, not accolades.

Take the Helm

- **Do the Quiet Work Well** – Take pride in institutional stewardship, policy refinement, command continuity, and unseen contributions that keep the Navy running.

- **Let Impact Speak, Not Ego** – Resist the urge to be seen doing the right thing. Do it consistently, and let its results validate your influence.

Eyes on the Horizon

Your greatest influence may never carry your name. Lead quietly, so the institution can carry on proudly.

CHAPTER 12:

ESTABLISHING YOUR LEGACY

Day 331 – Let Your Example Become Tradition

The Culture Code – Daniel Coyle

"Culture is not something you are. It's something you do."

Daniel Coyle dismantles the idea that culture is a fixed trait and instead shows how it is shaped by repeated, intentional behaviors. He writes that lasting cultures form when leaders model consistent actions that signal safety, purpose, and connection. These actions, even if small, compound into rituals that others adopt, not because they were told to, but because they resonate. At Pixar, Coyle highlights how creative leaders use structured feedback sessions like "BrainTrust" meetings where candor is encouraged, but delivered with empathy, shaping a culture of continuous improvement. He notes how these leaders deliberately model behaviors that prioritize trust, such as leaders admitting mistakes first during debriefs, which invites openness from the rest of the team. These consistent behaviors, practiced over time, become unspoken norms that define how the team operates proving that culture is formed by repeated choices that signal what matters. How you handle colors, how you open a debrief, how you walk the spaces and check on your team, these routines send a message. When done consistently and with authenticity, they stop being *yours* and start becoming *ours*. The best leaders create battle rhythms that eventually become tradition.

Take the Helm
- **Establish Meaningful Rituals** – Choose one positive ritual whether it's language, recognition, routine that reinforces your values and live it daily.

- **Lead Traditions That Outlast You** – Encourage your team to carry the most meaningful ones forward. Let your legacy become their tradition.

Eyes on the Horizon
Culture is built through repetition. What begins as your example becomes the team's tradition. Legacy forms when those you lead keep doing things the right way, long after you've left the space.

Day 332 – Legacy Is Built Daily

Legacy – James Kerr

"Leave the jersey in a better place."

James Kerr shows that enduring excellence comes from value-driven leadership and the willingness to serve something greater than oneself. The New Zealand All Blacks rugby team, known for their sustained dominance, live by the mantra, "Leave the jersey in a better place." Kerr explains that legacy is built not in victory parades or speeches, but in the small, consistent choices that leaders make. This principle resonates deeply in naval service. Whether you are a Petty Officer training your relief, a Department Head setting the tone, or a leading Seaman improving a watchstanding binder, your legacy does not rest on your name on the watchbill. The true legacy is the standard you leave for others to follow. Like the All Blacks, Sailors build legacy through mentorship, routine, and care. The leadership you model in maintenance meetings, the way you write and revise PQS, the culture you shape in your workspace, all of it becomes part of the ship long after you've checked out. We don't wear jerseys in the Navy, but every Sailor inherits the uniform and the responsibility to leave it, and the command, better than they found it.

Take the Helm
- **Be a Steward of Culture** – Identify one small, positive habit you can reinforce every day. Document it, model it, and pass it on.

- **Build What Lasts** – Build systems and mentorship relationships that don't depend on your presence. Create something your team will continue without you.

Eyes on the Horizon
Your legacy doesn't begin when you leave. It begins with what you lead today. Shape the culture now and it will carry forward long after you're gone.

Day 333 – Strengthen the Team by Teaching

The Culture Code – Daniel Coyle

"The road to success is paved with the idea that we learn faster when we learn together."

Daniel Coyle's research into high-performing groups reveals a key insight: cultures of excellence are built when people aren't just expected to learn, but also to teach. He explores how elite organizations build cohesion and mastery by reinforcing shared learning. At SEAL Team Six's training compound, for example, even the most experienced operators are tasked with teaching new teammates, not just to reinforce skills, but to deepen their own. This "teach-back" culture creates team resilience and mutual accountability. Nowhere is this more visible than in the Navy's warfare qualification programs. Whether earning your Surface, Submarine, Aviation, or Expeditionary pin, the expectation isn't just to learn, but also to teach. Once qualified, Sailors are expected to mentor and prepare the next wave of watchstanders. On ships, that culture is most visible in areas like the bridge, engineering and combat watchteam certifications, where experienced OODs, EOOWs, and TAOs are responsible not just for performance, but for replication. Each stage demands that newly qualified Sailors contribute by teaching and improving the team while reinforcing their own understanding. This is how high-performing commands thrive across turnovers and deployments. The strongest leaders aren't remembered for what they accomplished alone but for how many they equipped to carry the mission forward.

Take the Helm
- **Train with Repetition and Purpose** – Don't just show your relief what to do. Develop their judgment so they can lead others in turn.

- **Model the Long View** – Encourage your leaders to mentor their replacements. Legacy builds exponentially when leadership is taught to multiply.

Eyes on the Horizon
You haven't succeeded until someone else succeeds because of you. Teach leaders to teach and your influence will outlive your presence.

Day 334 – Write It Down, Pass It On

The Leader's Bookshelf – Admiral James Stavridis and R. Manning Ancell

"If it's not captured, it won't last."

Admiral James Stavridis and R. Manning Ancell underscore the simple truth that experience is only valuable if it's shared. Across interviews with more than 200 senior military leaders, one recurring theme stood out: leaders who took the time to write things down left a deeper, more lasting impact. Whether through personal journals, command guidance, reading lists, or continuity documents, written leadership leaves a legacy that spoken advice alone cannot match. "The best leaders read, reflect, and record," Stavridis notes. Capturing lessons in writing reflects both disciplined thought and a commitment to stewardship. If you want something to endure, you must codify it. This principle holds particular weight in the Navy, where turnover is constant and leadership is transient. Sailors come and go, but standard operating procedures, continuity folders, training guides, and mentorship notes can provide stability when people change. A well-written binder or personal leadership journal can help the next Sailor hit the ground running, reduce relearning, and maintain continuity of excellence. Whether you're leaving a division, department, or command, taking the time to document what worked and what didn't is a gift to the team that follows. Don't just leave memories. Leave behind a map others can follow.

Take the Helm

- **Document With Purpose** – Create or refine a continuity document before you depart that includes process notes, lessons learned, and leadership insights.

- **Make It Accessible** – Encourage your team to maintain a "living binder" like the "divo notebook," a centralized, up-to-date resource that tracks qualifications, training plans, watchbills, references, and daily routines.

Eyes on the Horizon

Legacy lives in what you leave behind. Write it down so the next watch starts where you finished.

Day 335 – Your Impact Is Measured in Stories

With the Old Breed – Eugene B. Sledge

"We carried the stories of the men we served with and they carried us."

Eugene B. Sledge recounts his harrowing experience as a Marine in the Pacific during World War II. The small acts of courage and humanity amid chaos became the stories that shaped him and those around him. He writes not only to document battles like Peleliu and Okinawa but to ensure that the courage, suffering, and brotherhood of the men he served with are not forgotten. Sledge reflects on moments like a Marine quietly breaking down after weeks of combat, or another showing resilience by cleaning his gear even as shells landed nearby. These accounts give voice to those who didn't write books or wear medals. By passing these memories on, Sledge affirms that storytelling is a moral act or an obligation to carry forward the truth and humanity of those who served beside him. Through his writing, he ensures that the impact of these men lives on, not in monuments, but in memory. You don't get to choose which stories your Sailors will carry but you do get to choose how you lead today. The legacy that truly lasts is more personal than formal. And it often begins in moments that feel small but are remembered as defining.

Take the Helm

- **Lead With Lasting Moments** – Understand that your smallest moments may become someone's most lasting lesson. Carry yourself accordingly.

- **Live the Legacy You Want Told** – Ask yourself, "What stories will they tell about me?" Then lead in a way that earns pride in the retelling.

Eyes on the Horizon
You'll live on in the stories they tell. Make your leadership worth remembering.

Day 336 – Lead So They Want to Stay

Leaders Eat Last – Simon Sinek

"When people feel safe, they stay. When they feel valued, they thrive."

Simon Sinek explains that great teams are built not through authority or incentive, but through trust, safety, and shared purpose. He writes, "Leadership is not about being in charge. It is about taking care of those in your charge." People don't stay in difficult jobs for the perks. They stay for the people. Sinek introduces the concept of the "Circle of Safety," where leaders protect and invest in their teams so they feel secure enough to take risks, grow, and commit. When individuals feel valued and supported, they are far more likely to remain and give their best not because they have to, but because they want to. This principle is foundational to retention and readiness. Sailors endure long hours, high demands, and significant personal sacrifice but they'll stay in commands where they feel respected, heard, and part of something meaningful. Leaders who walk the spaces, know their people's goals, and foster a sense of pride and belonging build environments that people want to return to. It doesn't mean lowering standards. It simply means leading with empathy, clarity, and trust. A Sailor who feels valued is a Sailor who invests back into the team. That sense of value becomes contagious, encouraging others to commit as well.

Take the Helm
- **Know Your Sailors** – Build personal connection with your Sailors. Learn their goals, challenges, and motivators.

- **Cultivate a Culture of Trust** – Create a culture of care without compromising standards. People stay where they grow and feel valued.

Eyes on the Horizon
Legacy lives in the crew that stays behind. Lead in a way that makes your Sailors want to stay in and stay sharp.

Day 337 – Mentor for the Moment After

The 7 Habits of Highly Effective People – Stephen R. Covey

"Begin with the end in mind."

Stephen R. Covey's second habit, begin with the end in mind, challenges leaders to act with long-term impact in view, urging individuals to clarify the destination before taking action. In leadership, this means mentoring not for dependence, but for independence. Nowhere is this more evident than in the Navy's enlisted leadership development programs. These programs are designed to shape Sailors who think critically, lead ethically, and make decisions independently when it counts. Chief-selects do not go through CPO 365 to become copies of their mentors. The goal is to equip them to uphold and advance the Navy's leadership tradition. Each phase builds from followership to self-awareness to deckplate command. Whether you're a Chief developing a new Work Center Supervisor or a Department Head preparing a Division Officer to take over, your influence multiplies when you train for autonomy. Mentorship is not about making a copy of yourself but about building someone ready to lead the moment you step away. When the end is part of your plan from the beginning, your departure becomes a handoff, not a disruption.

Take the Helm

- **Mentor for Absence** – Mentor your Sailors for moments you won't be there. Ask, "How would you lead this if I weren't here?"
- **Prioritize Judgment** – Focus mentorship on thinking, not just tasks. Teach judgment, not just execution.

Eyes on the Horizon
Mentorship should extend beyond your time in the seat and into the next chapter of your Sailors' careers. Build them for what's next, not just what's now.

Day 338 – Build Reputation Through Reliability

It's Your Ship – D. Michael Abrashoff

"The most important thing a captain can do is to see the ship through the eyes of the crew."

While leading *USS Benfold*, Captain D. Michael Abrashoff's leadership became synonymous with reliability because of his presence. One striking example occurred when a critical systems inspection revealed deficiencies in a key engineering space. Instead of passing blame down the chain, Abrashoff took full responsibility to the inspectors and used the failure as an opportunity to rally his crew. He didn't issue blanket orders or assign fault, instead he walked the space with his Sailors and asked them what had gone wrong. He listened and empowered junior Sailors to propose fixes and implemented their ideas. That empowerment only worked because they knew their captain would deliver on his end. Abrashoff ensured they had the tools, time, and command backing to carry out improvements. He eliminated unnecessary administrative burdens, streamlined communication, and gave junior Sailors ownership of the solution. This outcome built lasting trust. The Sailors saw that their input mattered. Over time, this led to *USS Benfold* becoming one of the Navy's top-performing ships, with improved retention, operational readiness, and morale. Abrashoff's reputation wasn't built on one act, but on a pattern: when he said he'd support his team, he meant it and delivered.

Take the Helm
- **Keep Every Commitment** – Follow through on commitments. If you say it, do it. Let your reputation be grounded in your consistency.
- **Reinforce Reliability as a Standard** – Teach your Sailors that reliability is leadership currency being counted on is more valuable than being noticed.

Eyes on the Horizon
Legacy is built on consistency. Show up, deliver, and repeat, and your Sailors will follow your example with trust.

Day 339 – Share the Credit, Own the Weight

Call Sign Chaos – Jim Mattis and Bing West

"A leader must take responsibility for everything while giving away the glory."

One of Jim Mattis' and Bing West's core tenets of leadership is that responsibility flows uphill and credit flows down. They believed deeply in elevating others and insisted that praise be directed toward those executing the mission. Mattis blended humility and accountability to create enduring trust and influence, living his principles across every echelon of command. When asked about his unit's successes, he deflected the praise to his subordinates, stating that, "it was the NCOs who made it happen." Conversely, when things went wrong, Mattis never pointed down. He took the weight publicly, even when the fault lay elsewhere. During the early stages of the Iraq War, as a Division Commander, he made it clear when he said, "I don't care who gets the credit, as long as the mission gets done." Naval leaders should follow this model in everything from evaluations to quarters, from awards to after-action reviews. When leaders recognize the unsung Sailor during quarters and stand tall when the mission falls short, they reinforce a culture of accountability and humility. The team remembers that. Over time, your presence becomes synonymous with fairness and integrity, precisely because you never made the mission about you. That's leadership that lasts.

Take the Helm
- **Lift Others Publicly** – Publicly recognize your Sailors for team wins. Use your influence to elevate them.
- **Absorb Responsibility With Grace** – Shield your team when mistakes happen. Take the weight so they can focus on recovery and growth.

Eyes on the Horizon
Legacy is built when leaders lift others up and absorb the hit when necessary. Give away the credit. Keep the responsibility.

Day 340 – Let Them Write the Last Chapter

Leadershift – John C. Maxwell

"The greatest legacy a leader can leave is having developed other leaders."

John C. Maxwell shares the story of his own leadership transition at The John Maxwell Company, where he intentionally stepped back from day-to-day operations and elevated leaders like Mark Cole to CEO. He invested years mentoring Cole, giving him increasing responsibility, and allowing him to lead major initiatives while still under Maxwell's guidance. Eventually, Cole was making strategic decisions without needing approval. Maxwell highlights that his final "leadershift" wasn't about legacy through reputation, but legacy through reproduction: cultivating a leader who could not only sustain the mission but grow it independently. This gradual, deliberate transition is a textbook example of letting others write the final chapter with confidence and continuity. Leaders who hold too tightly may slow their team's growth. It often shows up in subtle ways: rewriting a junior Sailor's report instead of coaching them through it, dominating planning meetings instead of asking for input, or refusing to delegate key tasks out of fear they'll be done "wrong." Maxwell warns that this kind of control creates dependency rather than leadership. Let your Sailors brief the final inspection, run the last evolution, and finish strong without you calling the play. That's where real confidence is built. Because when your Sailors close the chapter on their own terms with pride and ownership, they carry your influence forward.

Take the Helm
- **Empower the Final Chapter** – In your final stretch, hand over space to your team. Let them finish the tour their way.

- **Prioritize Purpose Over Presence** – Step back with trust and reinforce belief in the mission. They'll rise if they know why it matters.

Eyes on the Horizon
Your legacy is strongest when others take ownership of it. Step back and let your Sailors write the final pages with pride.

Day 341 – Lead the Transition

The Mission, the Men, and Me – Pete Blaber

"Smart leaders don't cling to control. They shape the handoff."

Pete Blaber writes that leadership isn't just about commanding through action, but about shaping outcomes through preparation, adaptability, and foresight. One of his core principles is that the mission always comes first, and that includes the mission of a smooth leadership transition. Wise leaders recognize that lasting influence shows in a team's continued strength after their departure. Prioritizing continuity and empowering the next leaders keeps momentum alive. Strong teams carry that drive forward. Across all military branches, transitions happen often. Commanding Officers, Division Officers, Chiefs, and Work Center Supervisors all rotate out, some after years of shaping a team. The challenge is not just finishing strong but handing off with intentionality. Leaders who prepare well don't simply turn over paperwork. They pass down context, expectations, and culture. They help their teams stay anchored during change by clearly articulating shared values, maintaining standards, and giving the incoming leader a clear runway. The difference between a disrupted team and a thriving one often comes down to how the outgoing leader handled the transition not as an afterthought, but as part of the mission.

Take the Helm

- **Invest in the Handoff** – Invest in your turnover process. Leave behind knowledge, confidence, and momentum.

- **Anchor the Culture** – Reinforce the team's identity during transition. Remind them what they stand for, regardless of who's in charge.

Eyes on the Horizon

A leader's legacy takes shape in the standards and people who thrive beyond their watch. Pass the torch with the same purpose and pride that fueled your watch.

Day 342 – Make Space for the Next Voice

Daring Greatly – Brené Brown

"Sometimes the most courageous thing you can do is step back."

Brené Brown redefines vulnerability as a powerful form of leadership. She highlights a school principal who transformed her leadership approach by creating space for others to speak and lead. Recognizing that faculty meetings had become one-sided, she began each session by asking, "What's one thing you're struggling with, and how can we support you?" This small but intentional act shifted the culture from top-down control to collaborative leadership. Teachers began voicing challenges, sharing ideas, and supporting one another, ultimately improving morale and performance. Brown uses this example to show that stepping back isn't relinquishing leadership but instead inviting others to step forward and shape the environment alongside you. Commands that run effective Sailor 360 programs often place junior Sailors in the forefront, not just to participate, but to lead. These Sailors facilitate discussions, guide peer learning, lead PT sessions, and brief command leadership. By intentionally creating space for emerging voices, Sailor 360 fosters grassroots leadership, reinforces trust, and reshapes command culture from the deckplates up. The result is professional growth, as well as a culture of empowerment. Giving junior leaders the space to make calls, brief the team, or solve problems their way is a sign of trust. And in doing so, leaders ensure that their legacy lives on in the voices they helped rise.

Take the Helm

- **Empower the Next Leader** – Identify a leader ready to step up. Trust them with real authority while you're still in the fight so they're already leading when you transfer out.

- **Let Go With Grace** – Support your replacement publicly. Model respect, trust, and transition maturity.

Eyes on the Horizon

The most confident leaders step aside at the right time. Let others rise, and your influence will rise with them.

Day 343 – Let Your Standards Speak for You

With the Old Breed – Eugene B. Sledge

"War is brutish, inglorious, and a terrible waste. But the worst part is how quickly it strips men of their standards, unless those standards are deeply ingrained."

Eugene B. Sledge describes the unimaginable conditions of heat, disease, exhaustion, and death faced by Marines on Peleliu and Okinawa. Yet what stood out most was not just the violence, but the discipline that separated those who held the line from those who lost themselves. He writes, "The men who endured were those who didn't surrender their standards even when no one was watching." In leadership, especially in warfighting environments, it's easy to assume that toughness is about aggression. But real toughness is consistency, repacking the same gear, cleaning the same weapon, running the same drill, especially when you don't feel like it. Navy leaders face their own version of this daily. They must enforce watchstanding discipline, procedural compliance, and uniform standards when it's uncomfortable, unpopular, or inconvenient. Your Sailors don't need a slogan. They need proof. And that proof shows up in the tone you set when no one's looking. Every unchecked detail becomes permission. Every quiet correction becomes culture. The leaders who leave a mark aren't the loudest, they're the ones whose standards echo through generations of Sailors who inherit their expectations.

Take the Helm

- **Set a Non-Negotiable** – Identify one standard in your space you won't compromise. Make it part of your identity.

- **Lead Through Consistency** – Coach your junior leaders to correct with calm consistency. Legacy grows through what you protect.

Eyes on the Horizon

Your standard will outlive your rank. Enforce it quietly, daily, and it will speak long after you're gone.

Day 344 – Don't Just Leave a System; Leave a Spirit

Turn the Ship Around! – L. David Marquet

"The goal is not just to build a better system but to build better people who can build better systems."

L. David Marquet challenges the traditional view of leadership as control and direction. Instead, he advocates for a model where leaders create more leaders where authority is pushed to the edge and every Sailor is expected to think, decide, and act. Marquet's core insight is that a lasting system must be regenerative. Nowhere is this more evident than during high-stakes inspections like INSURV, Supply Management Certification, Engineering and Combat Systems Light Off Assessments. These are some of the Navy's most demanding assessments. They're rigorous, sometimes no-notice evaluations that test a command's material readiness, supply systems, engineering safety, and combat capability. The strongest commands are the ones whose junior Sailors already know the standards, understand their role in success, and have been trained by Chiefs and LPOs who mentored with Marquet's regenerative model in mind. When the spirit of initiative outlasts your presence, that's true leadership. Leaders are often praised for programs or readiness gains achieved during their tour, but the deeper test is what happens after they leave. If the system collapses or stalls, then it was never sustainable, but if your Sailors continue adapting, improving, and holding the line, then you've built something far more powerful: a culture of empowerment.

Take the Helm

- **Foster Ownership Daily** – Challenge your team to improve the systems they inherit not just maintain them. Leave behind people who innovate, not just follow.

- **Build Leaders, Not Dependents** – Measure your success by what your team creates after you leave not just what you created for them.

Eyes on the Horizon

Processes may fade, but people who keep building define a leader's true legacy. Inspire their spirit, and your influence will echo forward.

Day 345 – Inspire by Living the Example

Helmet for My Pillow – Robert Leckie

"A man doesn't forget the leader who walked the same mud he did."

Marine veteran Robert Leckie recounts the brutal conditions of the Pacific campaign and the leaders who left a lasting impact. He vividly recounts a moment during the Battle of Peleliu when, under relentless shellfire and stifling heat, one of his sergeants crawled forward, not to give orders, but to dig in beside the men, sharing rations and reinforcing their position. That small act of presence and solidarity cut through the chaos more than any command could. This principle holds true across the fleet today. It's not the policy brief that builds respect, but the example Sailors see day after day. When leaders show up for fresh water washdowns, line up for FOD walkdowns, or shoulder a mop during cleaning stations, they prove that no task is beneath them and no standard is beneath their enforcement. These moments speak louder than any all-hands message. They tell your Sailors: I don't just expect the standard. I live it beside you. You don't have to give orders to shape a culture. You just have to model the one you expect. As Leckie's story reminds us, when your Sailors see your boots in the same grind they're in, they'll follow you. Not because they have to but because they trust you've earned the right to lead.

Take the Helm

- **Live the Lesson First** – Identify one small action you want your Sailors to adopt then live it with consistency and without needing credit.

- **Lead Before You Speak** – Reinforce that leadership starts before the orders are given. Your behavior is always briefing the team.

Eyes on the Horizon

Your influence begins before you speak. Show them what right looks like, and they'll carry that image forward.

Day 346 – Turn Mentorship Into Momentum

The 12 Rules for Life – Jordan B. Peterson

"Encouragement is not optional. It is leadership."

Let your mentorship create forward motion. Jordan B. Peterson emphasizes that people are capable of extraordinary growth but often only after someone else sees it in them first. He writes, "Treat people as if they were what they ought to be and you help them become what they are capable of being." This principle comes to life in the simple but powerful act of Commanding Officer (CO) recognition. During all-hands quarters or post-drill debriefs, when a CO calls out a junior Sailor for calm leadership during a casualty or recognizes unexpected initiative in front of peers, it sends a message that, "We see you. You matter." Those moments often create momentum that lasts far beyond the praise itself. When done with intention, a single word of encouragement can propel a Sailor toward greater confidence, performance, and leadership setting a course they might not have chosen alone. Leaders across all levels play a critical role in making that happen. LPOs, LCPOs, and Division Officers have the power to elevate their Sailors by recommending them for formal recognition. Whether it's a name passed up for all-hands acknowledgment or a note to the Chief's Mess or Wardroom, advocating for your Sailor costs nothing but it can change everything. When you choose to call out potential, you give someone permission to see themselves differently and that can be the spark that changes everything.

Take the Helm
- **Speak to Potential** – Take time to identify and tell Sailors what you see in them. Be the first one who believes out loud.

- **Encourage With Purpose** – Include encouragement in every critique. Build futures, not just files.

Eyes on the Horizon
Mentorship is belief in motion, not just advice. Give your Sailors momentum and they'll carry it farther than you can imagine.

Day 347 – Legacy Is the Culture You Leave Behind

Helmet for My Pillow – Robert Leckie

"They were gone, but their presence remained with us, in the way we spoke, the way we fought, and the way we carried ourselves."

As a Marine in the Pacific, Robert Leckie witnessed firsthand how the behaviors, attitudes, and quiet leadership of certain men outlasted the individuals themselves. He describes Marines like "Hoosier," a quiet, disciplined, and deeply loyal rifleman whose calm courage and unspoken loyalty shaped the behavior of those around him. That reputation followed Hoosier ashore in the Pacific, where his unspoken leadership, steady movements, measured speech, and careful packing became a model for his fellow Marines. When Hoosier was killed in action, his legacy didn't end. Leckie and others still carried themselves in his image, packing their gear the way he did, moving through patrols with his discipline, and speaking with the same measured tone. Hoosier's example became the unwritten culture, shaping how Marines fought, spoke, and held themselves under fire. This simple anecdote is proof that true leadership endures through influence, not position. That's the essence of cultural legacy. Leckie's account reminds us that even in the most chaotic and unforgiving environments, a legacy of perspective, of how to think, act, and endure, can be one of the most powerful gifts a leader leaves behind.

Take the Helm
- **Model Enduring Behaviors** – Identify the behaviors you want to endure. Model them until your team adopts them as their own.

- **Lead Through Culture** – Reinforce culture daily in small acts. Language, expectations, and rhythms define what remains.

Eyes on the Horizon
The truest legacy is the culture that outlives you. Build one your crew will be proud to protect.

Day 348 – Your Voice Becomes Their Inner Voice

Fearless – Eric Blehm

"The leaders we remember live on in our decisions even when they're not around."

Eric Blehm tells the story of Navy SEAL Adam Brown, not just through his actions, but through the lasting impact of his character and leadership. Those who served with Brown recalled how he spoke, how he carried himself, and how his presence shaped their thinking. One SEAL remembered how Brown would quietly say, "We have this," just before breaching a door. Never loud, never boastful, just steady and certain. That calm, confident phrase stuck with his teammates, who later found themselves repeating it silently in their own missions long after Brown was gone. Whether through steady words in chaos or encouragement during hardship, a leader's tone often becomes the framework through which others learn to lead themselves. Every word you speak in a moment of pressure, every calm correction, and every phrase you repeat during drills or debriefs can become part of your team's inner compass. Sailors don't just remember what you said, they remember how you said it and how it made them feel in the moment they needed it most. Over time, these patterns become instinctive. The best leaders understand that their influence isn't limited to their presence. Real impact continues in the thoughts, habits, and decisions of those they've shaped. Your voice, if used with care and purpose, can live in every critical decision your team makes even after you're gone.

Take the Helm
- **Speak With Purpose in Pressure** – Be mindful of how you speak in moments of pressure. Your words may become someone's guide later.

- **Reinforce Repeatable Phrasing** – Build a leadership rhythm of calm, focused phrasing. Those patterns will outlive your presence.

Eyes on the Horizon
Leadership doesn't end at your transfer. Make your voice one they can lead even after you're gone.

Day 349 – Legacy Is a Leadership Tree

Leading With the Heart – Mike Krzyzewski

"A leader's job is not to put greatness into people, but to recognize that it already exists and to create the environment where that greatness can emerge and grow."

Coach K emphasized personal connection, listening with purpose, and creating a team environment where players could hold each other accountable. His legacy was not the number of games won, but how many of his players went on to become leaders themselves in coaching, business, the military, and beyond. One standout example is his mentorship of Jon Scheyer, who would go from player to assistant to head coach at Duke. Coach K didn't just prepare Scheyer to win games, but rather he prepared him to lead a legacy. What truly measures legacy in the Navy isn't medals or sea time, but the leaders who stand ready because of you. Every Sailor you develop, mentor, and trust with responsibility becomes a branch on your leadership tree. When you invest in a junior Sailor's confidence, advocate for their advancement, or challenge them to take charge of a brief or evolution, you create the environment where their potential can emerge. That investment pays forward when they step up for others, train their reliefs, and build a strong division without you, your influence endures. A legacy of service means shaping a culture that outlives your watch. You're not just leading today. The example you set shapes the leaders who will define tomorrow.

Take the Helm

- **Grow Leaders, Not Just Performers** – Focus on growing leaders, not just performers. Develop judgment, ownership, and mentorship skills.

- **Invest in Multipliers** – Track your impact by watching others succeed without you and invest more in those who will pass it on.

Eyes on the Horizon

The strongest legacy is one that keeps expanding. Build leaders who build leaders and your impact becomes exponential.

Day 350 – Leadership Is an Act of Multiplication

Multipliers – Liz Wiseman

"The best leaders make everyone around them smarter."

The mark of a good leader is a team that feels valued enough to offer their best, every time. Liz Wiseman illustrates this principle through the example of Lutz Ziob, former general manager of Microsoft Learning. Rather than dictating direction, Ziob empowered his team to identify key opportunities and own their outcomes. When faced with the challenge of redefining Microsoft's certification programs, he invited a range of voices into the decision-making process, junior staff included, believing their insights would shape a better product. This open, participative approach not only delivered better results, it also developed leadership capabilities across the team. Wiseman highlights how Ziob's multiplier mindset helped create a culture where team members felt intellectually challenged, personally trusted, and professionally elevated, all hallmarks of legacy-building leadership. The multiplier effect is essential for building resilient, self-sustaining teams that can operate confidently without constant oversight. A Division Officer who trains their LPOs to run quarters, manage qualification boards, and brief the Department Head isn't giving up authority, but instead building leadership capacity. Similarly, Chiefs who empower junior Sailors to lead a working party or mentor others through watch qualifications are planting the seeds of ownership and pride. Leadership as an act of multiplication means stepping back, not to disappear, but to create space for others to step forward.

Take the Helm

- **Delegate Growth** – Assign junior Sailors leadership tasks with real impact. Let them grow through responsibility.

- **Ask, Don't Just Tell** – Use questions to develop thinking, not just compliance. Challenge Sailors to own solutions.

Eyes on the Horizon

Multiplying leadership means your legacy expands every time someone you empowered empowers someone else. When your Sailors leave stronger than they arrived, you've done more than lead. You've set up success for the future generation.

Day 351 – Set the Tone, Then Step Back

Legacy – James Kerr

"Leaders don't create followers. They create more leaders."

James Kerr describes how head coach Graham Henry transformed his leadership style during his tenure with the All Blacks, shifting from a command-and-control approach to one grounded in trust and distributed leadership. Early in his career, Henry was known for his tightly controlled methods, but after a disappointing performance at the 2007 Rugby World Cup, he recognized the need for a new model. Instead of directing every decision, he empowered a senior leadership group within the team to take ownership of standards, discipline, and preparation. This included veterans like Richie McCaw and Dan Carter, who became extensions of the coaching philosophy on the field and in the locker room. Henry intentionally stepped back from day-to-day micromanagement, trusting his players to hold each other accountable and lead from within. This approach built long-term resilience into the team culture, ensuring that leadership was embedded at every level, not reliant on the coach's presence to function. A departing Chief who empowers their LPOs to lead watchbills, enforce standards, and train junior Sailors without constant oversight sets a tone that survives turnover. A Department Head who involves JOs in planning boards, inspections, and briefings builds ownership. When you step back and they step forward with confidence, that's when you know your leadership truly endures.

Take the Helm

- **Reinforce Before You Depart** – Use your final month to cement your team's belief in themselves not in you.

- **Empower Future Voices** – Quiet your presence so their voices can rise. Your shadow should shelter, not dominate.

Eyes on the Horizon

Great leaders finish strong by letting go. End with clarity, trust, and a steady handoff.

Day 352 – Be Remembered for What You Gave

Daring Greatly – Brené Brown

"At the end of our lives, we won't be judged by our titles but by how deeply we cared."

Raised in rural South Carolina and of Cherokee descent, Boatswain's Mate First Class James E. Williams joined the Navy at 16 and served with distinction for 20 years, becoming one of the most decorated enlisted Sailors in U.S. history. His defining moment came on Oct. 31, 1966, while commanding Patrol Boat River 105 in the Mekong Delta. Surrounded and outgunned by enemy forces, Williams made bold tactical decisions over a three-hour firefight, coordinating air support and returning fire with unrelenting precision. His courage and calm under fire saved lives and turned a likely defeat into a decisive victory. But Williams' legacy was not just his battlefield heroism, but rather it was the example he set every day: leading from the front, mastering his craft, and making junior Sailors feel seen, capable, and valued. He was known for mentoring his crew, walking the deckplates, and ensuring every Sailor understood both their worth and their responsibility. Today's Sailor must balance readiness with resilience and accountability with empathy. Williams showed how to do all of that while never losing sight of the mission or the people. His legacy lives on because he lived the standard with consistency, courage, and care. That's the kind of leadership that echoes through generations and shapes the Navy's future. One example at a time.

Take the Helm
- **Practice Intentional Care** – Find one way each week to show your Sailors they matter. Your care may become their reason to keep going.

- **Lead With Heart and Strength** – Balance toughness with humanity. Legacy isn't built from pressure alone, but from presence.

Eyes on the Horizon
Lead with heart and your legacy will lead long after you.

Day 353 – Build a Legacy Bigger Than Your Name

Start With Why – Simon Sinek

"The goal is not to be the hero but to inspire others to become one."

Simon Sinek exemplifies this principle through the story of the Wright brothers, who succeeded in powered flight not because of fame or funding, but because of their deeply held belief in advancing human progress. Unlike their better-funded rival Samuel Langley, who chased recognition, the Wright brothers inspired their small team with purpose instead of ego. Sinek explains that this clarity of *why* is what allowed them to persist through failure and innovate together. Their example created generations of innovators, a ripple effect that long outlasted their own names. The Navy's core values of Honor, Courage, and Commitment also offers a clear example of this principle. Ownership of these values never rested with one person or command. Their purpose was to guide generations upon generations of Sailors long after their names faded. That is legacy beyond identity. Whether you're leading a division, a work center, or an entire command, your goal is not to be remembered as the centerpiece but to build upon others, something that functions with integrity and pride long after you're gone. As Sinek reminds us, the goal is not to be the hero, but to inspire others to become one.

Take the Helm
- **Teach the Meaning, Not Just the Method** – Communicate the *why* behind your expectations, corrections, and mentorship. Make your values visible and transferable.

- **Cultivate Lasting Ownership** – Tie your leadership to something bigger than yourself like heritage, mission, and team identity, so your legacy can grow.

Eyes on the Horizon
Great leaders seek continuation, not credit. Build something your Sailors believe in and your legacy will lead even when you no longer do.

Day 354 – Lead With a Mind Toward Memory

The Unforgiving Minute – Craig Mullaney

"You have one minute to prove who you are and sometimes that's all you'll get."

Craig Mullaney reflects on the reality that leadership is often remembered not in broad terms, but in singular, high-stakes moments. One decision. One reaction. One minute. You don't always get to choose the moment that defines you, but you do get to prepare for it every day. As a young platoon leader in Afghanistan, Mullaney experienced this first-hand during a firefight that erupted unexpectedly while his team was clearing a village. One of his Soldiers was hit. In that moment, Mullaney didn't have time to plan, only to lead. He called for suppressive fire, coordinated a medevac, and physically helped carry the wounded Soldier to safety. How he responded and who he was under pressure defined his relationship with the platoon from that day forward. He later wrote that despite months of training and preparation, it came down to "the unforgiving minute," the span of time where judgment, courage, and presence become more important than any title. These are the moments that happen during a man overboard call or a sudden Sailor emergency. Your reaction is what will be remembered. You don't choose when your defining moment arrives. But you do get to choose how you prepare for it day after day.

Take the Helm
- **Lead With Intentional Presence** – Treat every decision, especially under stress, as part of your leadership legacy. Someone is always watching.

- **Prepare for the Defining Minute** – Lead each day with the awareness that your actions could become someone else's lasting memory. Train, speak, and decide as if today is the moment they'll remember you by.

Eyes on the Horizon
One moment can define your legacy. Lead each day like that moment is coming.

Day 355 – Lead Like It's Not About You

The War of Art – Steven Pressfield

"Resistance will tell you anything to keep you from doing your work. It will perjure, fabricate, it will seduce you. Resistance is insidious."

Steven Pressfield argues that ego is a form of "Resistance" that lures leaders into seeking approval over impact, visibility over value. Great leaders, like great artists, recognize that their role isn't to be the star but to steward something bigger than themselves. Pressfield shares his experience as a struggling writer working menial jobs like driving trucks and picking fruit while quietly committing to his craft. He emphasizes that turning pro meant abandoning the ego-driven need for recognition. "I didn't even tell people I was a writer," he confesses. "I was just doing the work." That humility, choosing discipline and devotion over performance and praise, illustrates his central point: ego seeks applause, but real leadership (and artistry) shows up regardless of validation. This might look like a Chief who cedes the spotlight so a junior Sailor can present the brief or a Division Officer who takes critique quietly and implements change without fanfare. When you lead like it's not about you, you remove yourself as the center of gravity and empower others to step into theirs. You won't always be remembered for what you said, but your Sailors will remember how you made them feel: capable, trusted, and ready. That's the kind of legacy ego can't touch.

Take the Helm

- **Lead for Outcome, Not Optics** – Check your motivation during high-visibility moments. Are you leading for optics or for outcome?

- **Practice Humble Leadership** – Model servant leadership by coaching without credit and making others feel seen, heard, and capable.

Eyes on the Horizon

They may not remember your title, but they'll remember how you made them better. Lead for them, not for you.

Day 356 – Leave the Team Better Than You Found It

Good to Great – Jim Collins

"The signature of a great leader is what happens after they're gone."

Jim Collins explains that "Level 5 leaders" combine personal humility with fierce resolve laying a foundation others can build on. He illustrates this principle through the example of Darwin E. Smith, the humble CEO of Kimberly-Clark who led the company's transformation from a struggling paper producer into an industry leader outperforming Procter & Gamble. Rather than chasing headlines or personal credit, Smith made bold, long-term decisions like selling the company's core paper mills to invest in consumer products like Kleenex and Huggies. These moves were initially criticized, but over time proved visionary. Importantly, Collins highlights that Smith built a culture and leadership team that sustained success well after his departure, exemplifying how legacy is measured by continuity, not charisma. This approach of empowering successors and anchoring the mission in principles rather than personality is what Collins defines as Level 5 leadership: quiet, disciplined, and focused on what endures. This means leaving behind a division that doesn't stall when you check out, a training program that continues to sharpen Sailors, and a culture of professionalism that survives turnover. When you prepare others to lead, empower your Sailors and document your playbook so others can build on it. That's the mark of a leader whose influence outlasts any tour: one who leaves behind not just results, but ready leaders.

Take the Helm

- **Lead for Succession** – Prepare your team to operate without you. Train, document, and empower early and often.

- **Create Enduring Systems** – Ask yourself: "Will they thrive when I'm gone?" Then lead until the answer is yes.

Eyes on the Horizon

You haven't succeeded until your absence proves your leadership. Build it to outlast you and it will.

Day 357 – Inspire a Legacy, Not Just a Tour

The Infinite Game – Simon Sinek

"Leaders are not responsible for the results. They are responsible for the people who are responsible for the results."

Simon Sinek explains that finite-minded leaders chase quarterly metrics, award bullets, or tour-length wins. Infinite-minded leaders, by contrast, aim to leave something better than they found it: a culture, a standard, a spirit of growth. He illustrates this principle through the story of Bob Chapman, CEO of Barry-Wehmiller, who shifted his company's culture from one of short-term gains to long-term stewardship. Instead of laying off employees during economic downturns, Chapman invested in leadership development programs and human-centric management, not to boost quarterly performance, but to build an enduring culture of care, empowerment, and loyalty. By prioritizing people over profit and refusing to lay off workers during the 2008 financial crisis, Chapman deepened employee trust and loyalty. His approach led to lower turnover rates and sustained growth with Barry-Wehmiller expanding to over 100 acquisitions and generating more than $2 billion in revenue. This principle is especially relevant in the Navy, where leadership roles are time-bound, but their impact doesn't have to be. A Department Head who builds a strong training pipeline, a Chief who mentors their relief early and often, or a Division Officer who encourages junior Sailors to have a questioning attitude, all reflect infinite leadership. When you lead with a legacy mindset, your fingerprints remain, in the routines that persist, the trust that holds, and the Sailors who lead the next generation.

Take the Helm
- **Build Beyond Your Watch** – Set goals that will outlast your tour. Build systems and habits others can sustain.

- **Develop Depth** – Cross-train, mentor, and grow redundancy. Build a team that thrives without you.

Eyes on the Horizon
An infinite mindset multiplies. Meeting the standard is the start. Leaving a legacy is the goal.

Day 358 – Light the Path for Others

Legacy – James Kerr

"True leaders plant the seeds of trees whose shade they will never sit under."

James Kerr explores how the New Zealand All Blacks designed a culture that deliberately grows leaders at every level. One of the clearest illustrations of this is the "Captain's Run," the final practice before game day. It's not run by coaches but by the players themselves. This ritual is more transformational than ceremonial. It's a deliberate transition of ownership, where senior players step back and allow emerging leaders to set the tone, direct the pace, and lead the team. It's about passing the torch, not just hoping others are ready, but preparing them to lead by creating the space for them to do so. This example resonates deeply in the Navy, where the most enduring leaders build teams that evolve. When a Chief lets their LPO run the final pre-inspection brief or a Department Head gives a junior officer the reins during a high-visibility evolution, they're making a statement: "This is your watch now." Legacy in service is built by investing in people, so the mission carries on without your hand on the helm. True leadership doesn't aim to be remembered but rather aims to make others ready. When your Sailors step into the moment with confidence because of how you led them, that's legacy. You may never see where that path ends, but if you light it well, others will walk it boldly and extend it even farther.

Take the Helm
- **Coach for Long-Term Impact** – Identify at least one Sailor whose potential outpaces their opportunity. Coach, stretch, and trust them.
- **Teach the Why, Not Just the What** – Share the purpose behind your decisions. Teach vision, not just procedure.

Eyes on the Horizon
Your greatest legacy may come through someone else's leadership. Shine a light and show them the way forward.

Day 359 – Pass the Torch with Intention

The Leader's Bookshelf – Admiral James Stavridis and R. Manning Ancell

"Good leaders read. Great leaders write, teach, and mentor."

Leadership continues through education and elevation. Admiral James Stavridis and R. Manning Ancell profile senior military and civilian leaders to uncover a central truth: leadership is preserved not just through decisions but through deliberate transmission. The leaders he interviewed didn't view mentorship as extra work, but instead they saw it as a professional obligation. One specific example features their reflection on Admiral Mike Mullen, former Chairman of the Joint Chiefs of Staff. They note how Mullen prioritized mentorship and leader development during his career by hosting regular reading groups and encouraging officers to write and reflect. Mullen often assigned books to his staff and used them as a springboard for deeper conversations about leadership, ethics, and decision-making. Mullen actively created time to discuss the material and hear junior voices, reinforcing that mentorship isn't just about presence but about dialogue and shared growth. This approach shows how great leaders pass the torch with intention by investing in the intellectual and moral development of their successors. Whether through books, war stories, or one-on-one development, they viewed the act of teaching as critical to continuity and growth. From CPO 365 to Sailor 360 and officer training programs, the fleet thrives when seasoned leaders step back to coach. Leadership lives on through legacy, and legacy is built when we teach with intention.

Take the Helm

- **Develop the Leader, Not Just the Task** – Don't just supervise. Teach the reasoning behind actions so junior Sailors grow into future leaders.

- **Document and Share Lessons** – Keep a leadership journal or pass down your notes, lessons learned, and guidance to those stepping into your shoes.

Eyes on the Horizon

Legacy is who carries your example. Pass the torch clearly, so the flame of leadership burns brighter with every handoff.

Day 360 – Lead With Heart Even in the Fight

Leading With the Heart – Mike Krzyzewski

"People don't care how much you know until they know how much you care."

Coach K's success was built not just on strategy, but on deep, personal investment in his players. He wrote letters to parents, noticed emotional shifts in key players like Grant Hill, and supported alumni like Bobby Hurley long after their playing days ended. When the 2004 U.S. men's basketball team faltered internationally at the Summer Olympics in Athens, Coach K was selected to lead Team USA at the Beijing Olympics. He didn't lead with tactics. His strength came from the way he connected with others. Coach K met one-on-one with each one of his NBA stars to talk about purpose, service, and legacy. That care forged unity, and the result was a dominant Olympic gold in 2008. His leadership reminds us that loyalty is earned through presence, empathy, and belief in others. In the Navy, the same holds true. Sailors don't give their best to clipboard leaders. Genuine effort comes when leaders ask about their families, remember their goals, and invest beyond the Plan of the Day. Just like Coach K's players elevated their game because he invested in them as people, your Sailors will rise when they know their leader sees their worth. Care provides the backbone that allows discipline to endure when the stakes are highest.

Take the Helm
- **Lead Through Listening** – Learn what motivates your Sailors beyond the uniform. That knowledge builds lasting connection.

- **Show Up for the Person** – Celebrate wins, check in after loss, be present in the ordinary.

Eyes on the Horizon
Heart-centered leadership builds teams that believe in the mission. Rather than a distraction, care serves as the steady anchor that keeps teams strong.

PART IV:
CHARTING THE COURSE

Day 361 – Earned, Not Given

Make Your Bed – Admiral William H. McRaven

"If you can't do the little things right, you will never do the big things right."

This journey began with ownership. Before you could lead a watch team or inspire a crew, you had to first lead yourself with discipline, humility, and consistency. That's the foundation. Because wearing rank doesn't make you a leader. Showing up with integrity when unseen does. And every expectation you set for others traces back to the example you set for yourself. Admiral William H. McRaven learned this early in BUD/S. Each day began with a bed inspection because a perfectly made bed showed attention to detail, care for the task at hand, and a refusal to let standards slip. Those who failed at the small things like the bed inspection or uniform preparation or locker organization and gear checks were often the ones who faltered under pressure. McRaven makes it clear: the Sailor who can't be trusted to meet the standard in the quiet moments won't be trusted when the pressure rises. These small acts aren't meaningless. They're your resume. When you own them, consistently and without excuse, you lay the groundwork for trust. First with yourself, then with your team. Leadership is earned, not given, and it starts the moment you decide to lead yourself with purpose.

Take the Helm
- **Master the Small Things** – Take pride in daily responsibilities. Attention to detail and personal discipline reflect how you'll lead in crisis.

- **Lead by Quiet Example** – Be the Sailor whose consistency underwrites trust. Show your team who you are before you ever tell them.

Eyes on the Horizon
Great leadership begins in the quiet, unseen moments where discipline becomes habit and integrity becomes identity. Master the little things, and you'll become the foundation others trust when everything is on the line.

Day 362 – Live the Standard Every Day

The Unforgiving Minute – Craig Mullaney

"The only easy day was yesterday. Today, you must earn your place all over again."

As you close in on professionalism and moral leadership, the lesson becomes clear: your credibility doesn't rest on your ribbon rack, but rather on whether your actions match the standard you expect of others. Craig Mullaney learned this the hard way. As a young platoon leader in Afghanistan, he led with pride and confidence until the day one of his Soldiers was killed in a firefight. In the aftermath, Mullaney asked the questions all real leaders eventually face: Did I do enough? Did I train them right? Did I lead well when it mattered most? His reflection didn't offer easy answers, but it revealed the unrelenting nature of leadership. The standard isn't measured once. The standard is measured every day. In the hours of training before deployment, in the tone you set during a slow watch, in the way you correct errors, uphold dignity, and admit your own. Mullaney's story reminds us that leadership is about ownership. And that ownership must show up in your smallest behaviors long before it's tested in your biggest moments. The burden and gift of leadership is that *you* are the standard. Live the standard every day, not for applause, but because those who follow are learning how to lead by watching how you live.

Take the Helm

- **Pursue Leadership as Craft** – Treat each day as training. Prepare with purpose, mentor with care, model the standard, and keep sharpening your edge.

- **Stand on Principle, Not Popularity** – Let your choices reflect your values, not convenience. Integrity under pressure is the truest test of leadership.

Eyes on the Horizon
Your uniform may earn respect, but only your conduct will sustain it. Lead with daily discipline and let honor guide every choice.

Day 363 – Influence Is the Infrastructure

Leaders Eat Last – Simon Sinek

"The responsibility of leadership is not to come up with all the ideas but to create an environment in which great ideas can happen."

With the lessons behind you and the weight of influence within you, leadership is less about driving every outcome and more about creating the conditions where others can thrive, even when you're not in the room. That's what great leaders do, they build something that stands when they step away. Simon Sinek illustrates this through one of the Marine Corps' most powerful traditions: leaders eat last. There's no fanfare, no explanation, just a daily reminder that responsibility begins with service. That simple act communicates that their well-being matters more than your convenience. It's this kind of habitual selflessness that creates an environment where initiative grows, where trust becomes instinctive, and where teams operate not out of fear, but out of shared commitment. When Sailors feel safe, they hold themselves accountable. In every decision you've made, every tone you've set, you've been laying the foundation for something larger than yourself. That's because influence is the infrastructure: the unseen framework that holds the team together, the second you arrive to long after your tour ends and their mission continues.

Take the Helm
- **Build the Culture, Not Just the Schedule** – Set clear expectations, lead with respect, and shape a team that thrives with or without you on deck.

- **Lead Through Presence and Tone** – Be aware that how you carry yourself speaks louder than directives. Consistency and empathy under pressure are the bedrock of lasting influence.

Eyes on the Horizon
Teams don't rise to the level of their ambition. They rise to the strength of their cohesion and the clarity of your influence. Be the leader who builds the kind of environment where excellence is inevitable.

Day 364 – Lead With Composure, Decide With Courage

Leadership Strategy and Tactics – Jocko Willink

"Relax. Look around. Make a call."

You've learned that chaos is not an exception but rather a condition of command. The measure of a leader is not how they perform in calm waters, but how they respond when the fog rolls in and the pressure rises. Jocko Willink lived this in Ramadi, where the difference between life and death often came down to a single decision under extreme pressure. In one firefight, amid gunfire, confusion, and fractured comms, he paused, not out of indecision, but out of discipline. He took a breath. He detached. And then he led. That calm assessment allowed him to issue the right call and shift the outcome. But Willink reminds us this wasn't instinct but instead was trained composure. He built that habit not in combat, but in countless rehearsals, briefings, and honest debriefs. He carried that same discipline into peacetime, where mismanaged tempers and reactive leadership erode unity just as quickly as any misstep on the battlefield. Whether it's a fire in the plant, a man overboard, or a divisive meeting in the wardroom, the responsibility is the same: create clarity, maintain momentum, and move forward with conviction. In moments of pressure, your calm becomes the compass your team follows. Lead with composure, decide with courage, because that's how lasting trust is earned when the stakes are highest.

Take the Helm

- **Slow Down Under Pressure** – Don't let urgency override judgment. Breathe, assess, and act. Your calm will steady the storm.

- **Lead the Transition** – Anticipate resistance, communicate vision, and guide Sailors through change with empathy and confidence.

Eyes on the Horizon

Stress and change are constants, but confusion and panic don't have to be. In every uncertain moment, lead with purpose, and you'll turn turbulence into forward momentum.

Day 365 – Forge the Fighter, Shape the Thinker

Call Sign Chaos – Jim Mattis and Bing West

"If you haven't read hundreds of books, you're functionally illiterate. You can't coach and you can't lead."

Toughness alone is never enough. The warfighter's mindset demands both grit and clarity, a relentless drive to act with precision, and the discipline to think beyond the blast radius. Jim Mattis understood that chaos doesn't wait for calm, but rather rewards preparation. He carried history with him, literally and mentally, through every deployment. His footlocker held Thucydides, Clausewitz, Grant, and Churchill. His thoughts held centuries of judgment. Between operations, he read, wrote, and trained others to think. And when flawed orders came down, he didn't defer or deflect, instead he responded with clarity rooted in precedent, offering refined alternatives that saved lives and preserved mission integrity. He anticipated where others reacted. He led through intellect, not impulse. This is the calling of a fully formed leader: to act with force, but not without thought. To step into the fight with a mind shaped by study, tempered by experience, and focused on more than the moment. The Navy teaches us to fight the ship, but first we must train the mind that will decide when and how. Forge the fighter through discipline and action, shape the thinker through study and reflection. True leadership lives in those who can do both decisively.

Take the Helm
- **Sharpen the Warfighter's Edge** – Train the body. Prepare the mind. Study the craft, learn from the past, and lead with clarity when the pressure hits.

- **Lead with Tactical Precision and Strategic Purpose** – Don't just react. Think. Every decision must serve both the mission and the moment.

Eyes on the Horizon
Toughness helps you survive the fight, but thoughtfulness helps you lead through it. The warfighter who reads, reflects, and prepares is the one others trust to bring them home.

Day 366 – Leave the Watch Better Than You Found It

The Leader's Bookshelf – Admiral James Stavridis and R. Manning Ancell

"The best leaders don't just inspire people to follow. They prepare them to lead."

This chapter completes the leadership journey: from learning to lead yourself, to shaping teams, to preparing the future. Admiral James Stavridis described counseling an officer torn between staying Navy or pursuing civilian service. Instead of pushing retention, he asked questions that clarified values, priorities, and purpose. That Sailor left the Navy but went on to lead a humanitarian organization shaped by the same character forged in uniform. Stavridis saw leadership not as control, but as stewardship. He passed down belief, not just instruction. And in doing so, he multiplied his influence without ever demanding it. Whether you share your story with a Sailor on the mess decks, mentor a rising leader at a crossroads, or simply show up when no one else does, you're creating legacy. You are now the one others watch, the one they quote, the one they'll remember when it's their turn to decide what kind of leader they want to be. That's what it means to leave the watch better than you found it.

Take the Helm

- **Prepare Others to Lead** – Use your experience to equip, empower, and elevate the next watch. That's how your leadership outlasts your time in uniform.

- **Stay in the Fight, Beyond the Fleet** – Lead in every arena: at home, in your community, and in your work. The uniform may come off but your influence never should.

Eyes on the Horizon

Your watch may be over, but your mission isn't. Legacy is the wake you leave in others. Make it steady, make it strong, and make it last.

I hope this book has accompanied you through a meaningful stretch of your leadership journey. So where do you go from here?

You keep sailing forward.

My hope is that you continue to refine your skills, sharpen your perspective, and return to these pages as a trusted companion for reflection and growth. Keep investing in the principle that leadership is a profession, one that demands daily intention, especially in naval service.

To support that journey, I created **helmandhorizon.org**, a platform filled with practical tools and daily leadership wisdom to help you lead with purpose and stay mission-ready. It is a resource harbor, for your next leg of your leadership journey, and I hope to cross paths with you there. Wherever your course takes you next, may it be steady, purposeful, and anchored in service.

Lead with purpose. Take the helm. Keep your eyes on the horizon.

WORKS CITED

Abrashoff, D. Michael. *It's Your Ship: Management Techniques from the Best Damn Ship in the Navy.* New York: Business Plus, 2002.

Ancell, R. Manning, and James Stavridis. *The Leader's Bookshelf.* Annapolis: Naval Institute Press, 2017.

Arbinger Institute, The. *Leadership and Self-Deception: Getting Out of the Box.* San Francisco: Berrett-Koehler Publishers, 2000.

Blehm, Eric. *Fearless: The Undaunted Courage and Ultimate Sacrifice of Navy SEAL Team SIX Operator Adam Brown.* Colorado Springs: WaterBrook Press, 2012.

Borneman, Walter R. *The Admirals: Nimitz, Halsey, Leahy, and King—The Five-Star Admirals Who Won the War at Sea.* New York: Little, Brown and Company, 2012.

Bradley, James. *Flyboys: A True Story of Courage.* New York: Little, Brown and Company, 2003.

Brown, Brené. *Dare to Lead: Brave Work. Tough Conversations. Whole Hearts.* New York: Random House, 2018.

Brown, Brené. *Daring Greatly: How the Courage to Be Vulnerable Transforms the Way We Live, Love, Parent, and Lead.* New York: Gotham Books, 2012.

Carnegie, Dale. *How to Win Friends and Influence People.* New York: Simon & Schuster, 1936.

Clear, James. *Atomic Habits: An Easy & Proven Way to Build Good Habits & Break Bad Ones.* New York: Avery, 2018.

Collins, Jim. *Good to Great: Why Some Companies Make the Leap... and Others Don't.* New York: HarperBusiness, 2001.

Collins, Jim, and Morten T. Hansen. *Great by Choice: Uncertainty, Chaos, and Luck—Why Some Thrive Despite Them All.* New York: HarperBusiness, 2011.

Covey, Stephen R. *The 7 Habits of Highly Effective People: Powerful Lessons in Personal Change.* New York: Free Press, 1989.

Dalio, Ray. *Principles: Life and Work*. New York: Simon & Schuster, 2017.

Duckworth, Angela. *Grit: The Power of Passion and Perseverance*. New York: Scribner, 2016.

Dweck, Carol S. *Mindset: The New Psychology of Success*. New York: Random House, 2006.

Frankl, Viktor E. *Man's Search for Meaning*. Boston: Beacon Press, 2006.

Goggins, David. *Can't Hurt Me: Master Your Mind and Defy the Odds*. Lioncrest Publishing, 2018.

Goleman, Daniel. *Emotional Intelligence: Why It Can Matter More Than IQ*. New York: Bantam Books, 1995.

Greene, Robert. *The 33 Strategies of War*. New York: Viking Penguin, 2006.

Greitens, Eric. *Resilience: Hard-Won Wisdom for Living a Better Life*. Boston: Houghton Mifflin Harcourt, 2015.

Harding, Richard, and Agustín Guimerá, eds. *Naval Leadership in the Atlantic World: The Age of Reform and Revolution, 1700–1850*. London: University of Westminster Press, 2017.

Holiday, Ryan. *Ego Is the Enemy*. New York: Portfolio, 2016.

Holiday, Ryan. *The Obstacle Is the Way: The Timeless Art of Turning Trials into Triumph*. New York: Portfolio, 2014.

Hone, Trent. *Learning War: The Evolution of Fighting Doctrine in the U.S. Navy, 1898–1945*. Annapolis: Naval Institute Press, 2018.

Hornfischer, James D. *Neptune's Inferno: The U.S. Navy at Guadalcanal*. New York: Bantam Books, 2011.

Hornfischer, James D. *The Last Stand of the Tin Can Sailors: The Extraordinary World War II Story of the U.S. Navy's Finest Hour*. New York: Bantam Books, 2004.

Kerr, James. *Legacy: What the All Blacks Can Teach Us About the Business of Life*. London: Constable, 2013.

Krzyzewski, Mike. *Leading with the Heart: Coach K's Successful Strategies for Basketball, Business, and Life*. New York: Warner Books, 2000.

Leckie, Robert. *Helmet for My Pillow: From Parris Island to the Pacific*. New York: Bantam Books, 1957.

Lencioni, Patrick. *The Five Dysfunctions of a Team: A Leadership Fable*. San Francisco: Jossey-Bass, 2002.

Liker, Jeffrey K. *The Toyota Way: 14 Management Principles from the World's Greatest Manufacturer*. New York: McGraw-Hill, 2004.

Marquet, L. David. *Turn the Ship Around!: A True Story of Turning Followers into Leaders*. New York: Portfolio, 2013.

Mattis, Jim, and Bing West. *Call Sign Chaos: Learning to Lead*. New York: Random House, 2019.

Maxwell, John C. *Leadershift: The 11 Essential Changes Every Leader Must Embrace*. New York: HarperCollins Leadership, 2019.

Maxwell, John C. *The 5 Levels of Leadership: Proven Steps to Maximize Your Potential*. New York: Center Street, 2011.

McChrystal, Stanley, Tantum Collins, David Silverman, and Chris Fussell. *Team of Teams: New Rules of Engagement for a Complex World*. New York: Portfolio, 2015.

McRaven, William H. *Make Your Bed: Little Things That Can Change Your Life... And Maybe the World*. New York: Grand Central Publishing, 2017.

Mullaney, Craig. *The Unforgiving Minute: A Soldier's Education*. New York: Penguin Press, 2009.

Patterson, Kerry, Joseph Grenny, Ron McMillan, and Al Switzler. *Crucial Conversations: Tools for Talking When Stakes Are High*. New York: McGraw-Hill, 2002.

Peterson, Jordan B. *12 Rules for Life: An Antidote to Chaos*. Toronto: Random House Canada, 2018.

Pressfield, Steven. *The War of Art: Break Through the Blocks and Win Your Inner Creative Battles*. New York: Black Irish Entertainment, 2002.

Ries, Eric. *The Lean Startup: How Today's Entrepreneurs Use Continuous Innovation to Create Radically Successful Businesses*. New York: Crown Business, 2011.

Roberts, Andrew. *Leadership in War: Essential Lessons from Those Who Made History.* New York: Viking, 2019.

Scott, Kim. *Radical Candor: Be a Kick-Ass Boss Without Losing Your Humanity.* New York: St. Martin's Press, 2017.

Sinek, Simon. *Leaders Eat Last: Why Some Teams Pull Together and Others Don't.* New York: Portfolio, 2014.

Sinek, Simon. *Start With Why: How Great Leaders Inspire Everyone to Take Action.* New York: Portfolio, 2009.

Sinek, Simon. *The Infinite Game.* New York: Portfolio, 2019.

Sledge, Eugene B. *With the Old Breed: At Peleliu and Okinawa.* New York: Presidio Press, 1981.

Stavridis, James. *Sea Power: The History and Geopolitics of the World's Oceans.* New York: Penguin Press, 2017.

Sun Tzu. *The Art of War.* Translated by Samuel B. Griffith. New York: Oxford University Press, 1963.

Toll, Ian W. *Six Frigates: The Epic History of the Founding of the U.S. Navy.* New York: W. W. Norton & Company, 2006.

Willink, Jocko. *Discipline Equals Freedom: Field Manual.* New York: St. Martin's Press, 2017.

Willink, Jocko. *Leadership Strategy and Tactics: Field Manual.* New York: St. Martin's Press, 2020.

Willink, Jocko, and Leif Babin. *Extreme Ownership: How U.S. Navy SEALs Lead and Win.* New York: St. Martin's Press, 2015.

Willink, Jocko, and Leif Babin. *The Dichotomy of Leadership: Balancing the Challenges of Extreme Ownership to Lead and Win.* New York: St. Martin's Press, 2018.

Wiseman, Liz. *Multipliers: How the Best Leaders Make Everyone Smarter.* New York: Harper Business, 2010.

ABOUT THE AUTHOR

Steven-Paul Lapid is a retired U.S. Naval Officer, leadership author, and military analyst with over 25 years of experience leading Sailors across surface warships, joint commands, and Naval Special Warfare units. Beginning his career as an enlisted Sailor, he steadily rose from the deckplates to commission as an officer, earning a reputation for his effective mentorship across all ranks and building cohesive, high-performing teams.

His operational tours included destroyers, cruisers, and amphibious warships, while his ashore assignments placed him in key roles at SEAL Team Three and across multiple global commands. In each assignment, he became a trusted voice in readiness, certification, and operational planning; recognized for his calm presence, tactical expertise, and practical approach to leadership.

He currently serves as the senior training and warfighting readiness policy analyst for Commander, Naval Surface Force, U.S. Pacific Fleet (COMNAVSURFPAC) where he drives policy development, readiness forecasting, and training optimization initiatives for the Navy's surface forces.

Steve is the founder of Helm & Horizon, a leadership development platform dedicated to inspiring Sailors through resources that blend timeless principles, naval history, and daily reflection. The initiative offers practical guidance drawn from his own service, contributions of fellow veterans, and the Navy's most storied figures.

He holds degrees from the University of Michigan and the University of Florida and is an alumnus of the Aspen Executive Seminar on Leadership, Values, and the Good Society. He lives with his family in La Jolla, California.

For more information and leadership resources visit helmandhorizon.org.